PIC Microcontroller
Project Book

PIC Microcontroller Project Book

For PICBasic and PICBasic Pro Compilers

John Iovine

Second Edition

McGraw-Hill

New York   Chicago   San Francisco   Lisbon   London   Madrid
Mexico City   Milan   New Delhi   San Juan   Seoul
Singapore   Sydney   Toronto

The McGraw-Hill Companies

Cataloging-in-Publication Data is on file with the Library of Congress

Copyright © 2004, by The McGraw-Hill Companies, Inc. All rights reserved. Printed in the United States of America. Except as permitted under the United States Copyright Act of 1976, no part of this publication may be reproduced or distributed in any form or by any means, or stored in a database or retrieval system, without the prior written permission of the publisher.

1 2 3 4 5 6 7 8 9 0 DOC/DOC 0 9 8 7 6 5 4

ISBN 0-07-143704-5

The sponsoring editor for this book was Judy Bass and the production supervisor was Sherri Souffrance. It was set in New Century Schoolbook by MacAllister Publishing Services. The art director for the cover was Margaret Webster-Shapiro.

Printed and bound by RR Donnelley.

McGraw-Hill books are available at special quantity discounts to use as premiums and sales promotions, or for use in corporate training programs. For more information, please write to the Director of Special Sales, McGraw-Hill Professional, Two Penn Plaza, New York, NY 10121-2298. Or contact your local bookstore.

 This book was printed on recycled, acid-free paper containing a minimum of 50% recycled, de-inked fiber.

Contents

Chapter 11 Speech Synthesizer 151

Microcontrollers

What Is a Microcontroller?

A microcontroller is an inexpensive single-chip computer. Single chip means that the entire computer system lies within the confines of the integrated circuit. The microcontroller existing on the encapsulated sliver of silicon has features and similarities to our standard personal computers. Primarily, the microcontroller is capable of storing and running a program, its most important feature.

The microcontroller contains a *central processing unit* (CPU), *random-access memory* (RAM), *read-only memory* (ROM), *electrically erasable programmable read-only memory* (EEPROM), *input/output* (I/O) lines, serial and parallel ports, timers, and other built-in peripherals, such as *analog-to-digital* (A/D) and *digital-to-analog* (D/A) converters.

Why Use a Microcontroller?

The microcontroller's ability to store and run unique programs makes it extremely versatile. For instance, a microcontroller can be programmed to make decisions and perform functions based on predetermined situations (I/O line logic) and selections. Its ability to perform math and logic functions allows it to mimic sophisticated logic and electronic circuits.

Other programs can make the microcontroller behave like a neural circuit or a fuzzy logic controller. Microcontrollers are responsible for the intelligence in most smart devices on the consumer market.

Microcontrollers Are the Future of Electronics

Look in any hobbyist electronics magazine and you will see articles that feature the use of microcontrollers either directly or embedded inside a circuit's design. Because of their versatility, they add a lot of power, control, and options for a small cost. It therefore becomes essential that the electronics engineer or hobbyist learns to program these microcontrollers in order to maintain a level of competence and to gain the advantages that microcontrollers can provide in their own circuit designs. In addition, if you examine consumer electronics, you will find microcontrollers embedded in just about everything.

Designer Computers—So Many Microcontrollers

A large variety of microcontrollers exist on the market today. We will focus on a few versatile microcontroller chips called PIC chips (or PICMicro chips) from Microchip Technologies.

The PIC Chip

Microchip's microcontrollers are commonly called PIC chips. Microchip uses *PIC* to describe its series of PIC microcontrollers. Although it is not specifically defined, the word *PIC* is generally assumed to mean programmable interface controller.

Better Than Any Stamp

The company Parallax sells an easy-to-use series of microcontroller circuits called the Basic Stamp. Their series of Basic Stamps (BS-1 and BS-2) uses Microchip's PIC microcontrollers. The Stamps are popular and easy to use because they are programmed using a simplified form of the Basic language. Basic-language programming is easy to learn and use. This programming is the Basic Stamps' main advantage over other microcontroller systems. Other microcontroller systems have a much longer learning curve because they force their users and developers to learn a niche assembly language—meaning a language specific to that company's microcontroller and no one else.

The Basic Stamp has become one of the most popular microcontrollers in use today. Again, the Basic Stamp's popularity (it bears repeating) is due to its easy-to-learn and easy-to-use Basic-language programming.

The Basic-language PIC compiler we will use to program the PIC chips uses a Basic language similar to the syntax used by the Basic Stamp series. Now you can enjoy the same easy language the Basic Stamp offers, along with two more very important benefits.

Benefit 1: Faster Speed

Our programmed PIC chips will run their program much faster than the Basic Stamps do. If we enter the identical Basic program into a Basic Stamp and into a PIC chip, the programmed PIC chip will run 20 to 100 times faster, depending upon the instructions used, than the Basic Stamp runs. Here's why.

The BS1 and BS2 systems use a serial EEPROM memory connected to the PIC chip to store their programs. The Basic commands in the program are stored as Basic tokens. Basic tokens are like short hand for Basic commands. When running the program, the Basic Stamp reads each instruction (token and data or address) over the serial line from the external EEPROM memory, interprets the token (converts the token to a *machine language* [ML] equivalent the PIC can understand), then performs the instruction, reads the next instruction, and so on. Each and every instruction goes through this serial load, read, interpret, and perform sequence for as long as the program runs. The serial-interface reading routine eats up gobs of a microcontroller's CPU time.

In contrast to this operation, when a PIC chip is programmed using the Basic compiler, the Basic program is first converted to an ML-equivalent program and saved as a file of hexadecimal numbers, appropriately called a hex file. The hex file is then uploaded directly into the onboard (EEPROM) memory of the PIC chip. Because the ML program (the hex file) is the native language of the PIC, the code does not need to be interpreted as it runs. It reads the ML program instructions directly from its onboard memory and performs the instruction. There is no external EEPROM or serial interface to an external EEPROM to eat up CPU time. This system enables our programmed PIC chips to run their code 20 to 100 times faster than the same Basic program code in a Basic Stamp.

Benefit 2: Lower Cost

The next factor is cost. Using the PIC chips directly will save you 75 percent of the cost of a comparable Basic Stamp. The retail price of the BS-1 that has 256 bytes of programmable memory is $34.00. The retail price for the BS-2 that has 2K of programmable memory is $49.00. The 16F84 PIC microcontroller featured in this book is more closely compared to the BS-2 Stamp. The 16F84 PIC chip we are using has 1K of programmable memory.

The 16F84 PIC chip's retail cost is $6.95. To this price add the cost of a timing crystal, a few caps, a few resistors, and a 7805 voltage regulator, all of which are needed to create the circuit equivalent to the Stamp. These components increase the total cost to about $9.00—still below 25 percent of the cost currently quoted for the BS-2.

This $9.00 PIC cost may also be cut substantially in some situations. The PIC 16F84 is a microcontroller with rewriteable, or Flash, memory. *Flash* is the term used to describe rewriteable memory. If, for instance, you designed a circuit (or product) for manufacture that doesn't need to be reprogrammed after its initial programming, you could use a *one-time programmable* (OTP) PIC microcontroller and save $2 to $3 on the microcontroller.

In any case, anyone who uses more than a few Stamps a year will find it well worth the investment to jump onto this faster and less-expensive microcontroller bandwagon.

If you are an experimenter, developer, or manufacturer, or plan to become one, the cost savings are too substantial to consider investing in any other system.

Bonus Advantage

The footprint of the 16F84 PIC microcontroller chip embedded in another circuit is smaller than the equivalent BS-2 Stamp. The reason is that the Stamps use an external serial EEPROM for memory. Although the BS-2 may at first glance look smaller, being contained in a 28-*pin dual inline package* (DIP), it is not necessarily. You can also purchase the 16F84 in a surface-mount style, and the resulting circuit will have a correspondingly smaller footprint.

PIC Programming Overview

Programming PIC microcontrollers is a simple three-step process: write the code, compile the code, and upload the code into a microcontroller. Following is an overview of the process; step-by-step instructions will be provided in the subsequent chapters.

Software and Hardware

You will need two items to begin programming and building microcontroller-based projects and robotics. First is the compiler, either the PICBasic Pro or PICBasic compiler, from microEngineering Labs, Inc. (see Figure 1-1). The PICBasic Pro compiler has a suggested retail price of $249.95. The PICBasic compiler has a suggested retail price of $99.95. In addition to a compiler, you also need the EPIC programming board and software, also from microEngineering Labs.[1] This package sells for $59.95 (see Figure 1-2).

FIGURE 1-1   PicBasic Pro and PICBasic software packages and manuals

[1]The names PICBasic Pro, PICBasic, and EPIC are trademarks of microEngineering Labs, Inc.

FIGURE 1-2 EPIC programming software and hardware

PICBasic and PICBasic Pro Compilers

The PICBasic and PICBasic Pro compilers both function in the same way. The program code, saved as a text file, is run through either the PICBasic or PICBasic Pro compiler. The compiler reads through the text file and creates (or compiles) an equivalent machine-code instruction listing (the hex file) of the program. The machine code is a list of hexadecimal numbers that represents the PICBasic program. The hex file is uploaded (or programmed) into the microcontroller. When the microcontroller is started, its CPU will run through the programmed list of hexadecimal numbers that run the PICBasic program. Uploading the machine code into the microcontroller is the job of the EPIC programmer board and software, which we will look at shortly.

The PICBasic Pro (Professional) compiler is considerably more expensive than the standard PICBasic compiler. The Pro version offers a richer, more enhanced Basic command syntax than is available in the PICBasic compiler package. A few of the additional commands that can be found in the Pro version allow the use of Interrupts, direct control of *liquid crystal display* (LCD) modules, *dual-tone multifrequency* (DTMF) out, and X-10 commands, to name a few.

While PICBasic Pro is a more sophisticated package, the compiler does not handle two of my favorite Basic commands—Peek and Poke. While the Pro manual lists the commands as functional, it emphasizes that "Peek and Poke should never be used in a PICBasic Pro program." There are easy work-arounds to using the Peek and Poke commands in the Pro version that will be covered when needed later on.

In the balance of this book, I will, at times, refer to both the PICBasic and PICBasic Pro compiler simply as *the compiler* or *the compilers*. This use saves me from continually writing PICBasic and PICBasic Pro compiler through out the book. When a distinction becomes necessary, I will call out the individual compiler.

The compiler program may be run manually in DOS or in an MS-DOS Prompt window. A third option, and one you will probably use, is running the compiler within a Windows program called CodeDesigner. CodeDesigner is discussed later in this chapter and more fully in Chapter 4.

The minimum system requirements for the compiler are an XT-class *personal computer* (PC) running DOS 3.3 or higher. The compiler can compile programs for a large variety of PIC microcontrollers.

EPIC Programmer

The second item needed to program a microcontroller is the EPIC programmer. The EPIC programmer consists of software (EPIC Program) and a programming carrier board (hardware). The EPIC software package has two executable files, one for DOS and another version for Windows.

It is the EPIC hardware and software that take the compiled .hex file generated by the compiler and upload it into the microcontroller where it may be run. The EPIC programmer is compatible with both the PICBasic and PICBasic Pro compilers.

The programming carrier board has a socket for inserting the PIC chip and connecting it to the computer, via the printer port, for programming (see Figure 1-3). The pro-

FIGURE 1-3   Close-up of EPIC programming carrier board

gramming board connects to the computer's printer port using a DB25 cable. If the computer has only one printer port with a printer connected to it, the printer must be temporarily disconnected when programming PIC chips. The EPIC programming carrier board supports a large variety of PIC microcontrollers.

Serial Port and *Universal Serial Bus* (USB) EPIC Programmer

The standard EPIC programmer connects to the PC parallel port. There is another style of EPIC programmer that connects to the PC's serial port. This serial-port programmer may also be connected to a USB port using a serial-to-USB adaptor.

Firmware

Many writers use the term *firmware* in reference to software that is embedded in a hardware device, which can be read and executed by the device but cannot be modified. So, when our program is embedded (or uploaded) into the microcontroller it may be referred to as firmware. Other phrases may include the term *firmware* instead of *software*, such as "upload the firmware" or "once the firmware has been installed into the device."

Consumables

Consumables are the electronic components—the PIC microcontroller chip itself and a few support components to get the microcontroller up and running. I recommend beginning with the 16F84 PIC microcontroller. The 16F84 is an 18-pin DIP chip with 13 I/O lines and has 1K×14 bits of rewriteable memory. The rewriteable memory allows you to reprogram the PIC chip up to 1,000 times to test and troubleshoot your programs and circuits. The minimal support components are a 5-*volt direct current* (VDC) power supply, an oscillator (4.0 MHz crystal) and one pull-up .25-watt resistor (4.7K ohm).

16F84 PIC Microcontroller

The PIC 16F84 microcontroller is shown in Figure 1-4. It is a versatile microcontroller with flash memory. The 1K×14-bit on-board flash memory can endure a minimum of 1,000 erase-write cycles, so you can reprogram and reuse the PIC chip at least 1,000 times. The program retention time, between erase-write cycles, is approximately 40 years.

Another feature of the 16F84 microcontroller is its programmable pins. The 18-pin chip devotes 13 pins to I/O. Each pin may be programmed individually for input or output. The pin status (I/O direction control) may be changed on the fly via programming. Other features include power on reset, power saving sleep mode, power-up timer, code protection, and more. Additional features and architecture details of the PIC 16F84 will be given as we continue.

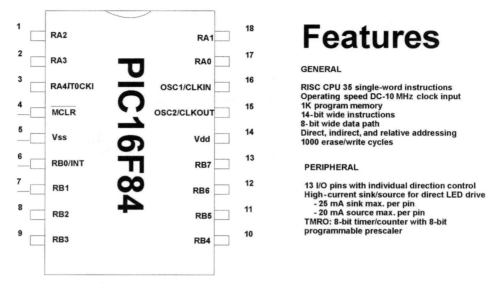

FIGURE 1-4   Pin-out of 16F84 PIC microcontroller *integrated circuit* (IC)

Step 1: Writing Code (The Basic Program)

Both the PICBasic and PICBasic Pro compilers are packaged with a free version of the CodeDesigner software. CodeDesigner is an *integrated development environment* (IDE) for writing and programming PIC microcontrollers. CodeDesigner is an advanced text editor capable of calling and using the PICBasic and PICBasic Pro compilers and the EPIC software.

If you don't want to use CodeDesigner, program text files may be written using any word processor, as long as it is able to save its text file as ASCII or DOS text. If you don't own a commercial word processor, you can use Windows Notepad, which is included with Windows 3.X, Windows 95/98, and XP/2000/NT. If you work at the DOS level you can use the EDIT program to write text files.

When you save the text file, save it with a .BAS suffix. For example, if you were saving a program named wink, save it as wink.bas.

Step 2: Using the Compiler

Once set up, the CodeDesigner software will call and control the compiler and programmer software. The compiler may also be run manually from a DOS window. To run the compiler program manually, enter the command *PBC* followed by the number of the

PIC chip being programmed (i.e., 16F84) and then the name of the source-code text file. For the PICBasic Pro compiler program the command starts with PBP instead of PBC, followed by the name of the source-code text file. For example, for the PICBasic compiler, if the source-code text file we created is named wink, then at the DOS command prompt enter the following:

```
PBC -p16f84 WINK.BAS
```

For the PICBasic Pro compiler, the command line would be:

```
PBP -p16f84 WINK.BAS
```

The compiler reads the text file and compiles two additional files, an .asm (assembly language) file and a .hex (hexadecimal) file.

The wink.asm file is the assembly-language equivalent to the Basic program. The wink.hex file is the machine code of the program written in hexadecimal numbers. It is the .hex file that is uploaded into the PIC chip.

If the compiler encounters errors when compiling the PICBasic source code, it will issue a list of errors it has found and will terminate. The errors listed need to be corrected in the source-code text file before it will successfully compile.

Step 3: Installing the Firmware, or Programming the PIC Chip

Connect the EPIC programming board to the computer's printer port using a DB25 cable. If you are using CodeDesigner, launch the EPIC programmer from the menu. The EPIC programming board must be connected to the parallel port and switched on before you start the software or the software will issue an error message: "EPIC programmer not found." Aside from the EPIC Windows software (EPICWIN.exe), which may be started manually in Windows or through the CodeDesigner software, there is also a DOS version of the program called EPIC.exe.

Figure 1-5 is a picture of the EPIC Windows program screen. Use the Open File option and select wink.hex from the files displayed in the dialog box. The file will load and numbers will be displayed in the code window on the left. Insert the 16F84 into the socket on the programming board and select the Program option from the Run menu. An alternative to using the menu option is to press the Ctrl and P buttons on the keyboard. The software is then uploaded into the PIC microcontroller and is ready to be inserted into your circuit and go to work.

Ready, Steady, Go

The following chapters contain step-by-step instructions for installing the software onto your hard drive and for programming your first PICmicro chip.

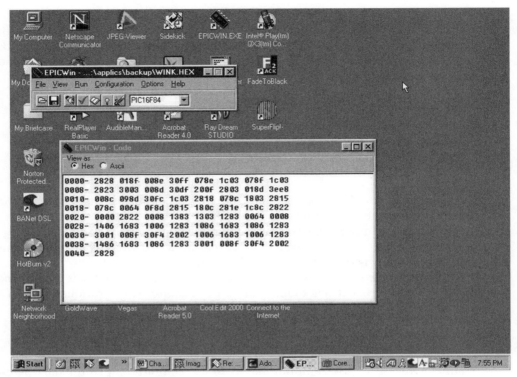

FIGURE 1-5   Windows version of EPIC software

Hardware and Software

The following is a list of the hardware and software needed to get going on programming microcontrollers and their retail prices.

PICBasic Pro Compiler	$249.95	or the PICBasic Compiler	$ 99.95
EPIC Programmer	$ 59.00		

Parts List

You will need the following parts for programming microcontrollers. You can find the parts at your local Radio Shack and they are also available from Images SI Inc., Jameco Electronics, and JDR MicroDevices. See the Suppliers Index for a complete listing.

16F84 microcontroller	$7.95
4.0 MHz Xtal	$2.50
Two 22 pF capacitors	
One Solderless Breadboard	Radio Shack PN# 276-175
One .1 uF capacitor	Radio Shack PN# 272-1069
Two Red LED's	Radio Shack PN# 276-208
Two 470 ohm resistors*	Radio Shack PN# 270-1115
One 4.7 K ohm resistor	Radio Shack PN# 271-1126
One 7805 Voltage regulator	Radio Shack PN# 276-1770
One 9-V battery clip	Radio Shack PN# 270 325

2

Installing the Compiler

To compile your Basic programs (text files) into data that can be uploaded into the PIC microcontrollers and run, you need to run the program text file through a compiler. The first step is to install the compiler software onto your computer's hard drive. The following instructions are for installing the PICBasic Compiler, and the instructions for installing the PICBasic Pro compiler are included as well.

Installing the PICBasic Compiler Software

First, create a subdirectory on your computer's hard drive for the PICBasic compiler software. In these instructions, I will use Windows Explorer (in Windows 95/98 /ME/2000/XP) to create this directory. Windows Explorer can be found in the Programs folder in Windows 95/98 (see Figure 2-1). For Windows ME/2000/XP users, Windows Explorer can be found in the Accessories folder (see Figure 2-2).

Next, create the subdirectory and name it PBC on the computer's hard drive. Copy all the files on the PicBasic diskette into the PBC subdirectory. For the conventions in this book, it is assumed that the reader's hard drive is drive letter C.

Now start the Windows Explorer program. Select your computer's hard drive (usually the C drive) in the Folders window and then go to the File menu and choose New. Click the Folder option (see Figure 2-3) and enter the name PBC in the New Folder icon (see Figure 2-4).

Place the 3.5-inch PicBasic compiler diskette into your computer's floppy drive, usually the A drive. Highlight the A drive in the Windows Explorer's Folder window, shown in Figure 2-5. All the files on the 3.5-inch diskette will be displayed on the right side. Select all the files, go to the Edit menu's options, and choose Copy (see Figure 2-6). Next, select the PBC directory on the left side of the Explorer window. Then go back to the Edit menu and select the Paste option. All the files and subdirectories on the 3.5-inch diskette will be copied into the PBC directory on the hard drive.

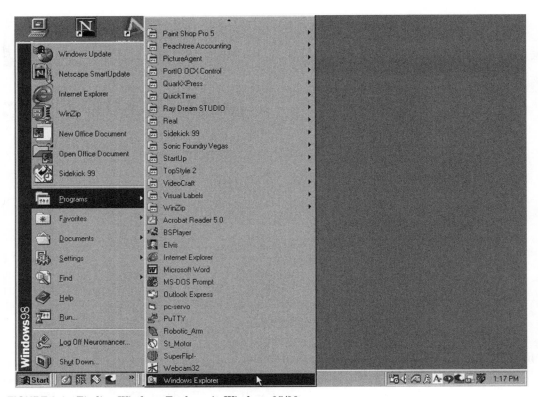

FIGURE 2-1   Finding Windows Explorer in Windows 95/98

An alternative to pasting the selected files is to select all the files as before, copy them, and then drag them into the PBC directory (see Figure 2-7).

Installing PICBasic Pro Compiler

Installing the PICBasic Pro compiler is not the same as the procedure outlined for the PICBasic compiler. To install the PICBasic Pro compiler, you must execute a self-extracting program that decompresses the necessary programs and files. It is recommended that you create a subdirectory named PBP on your computer's hard drive.

First, start the Windows Explorer program and select your computer's hard drive (usually the C drive) in the Folders window. Next, go to the File menu, choose New, and click on the Folder option (refer to Figure 2-3). Enter the name PBP in the New Folder icon (refer to Figure 2-4).

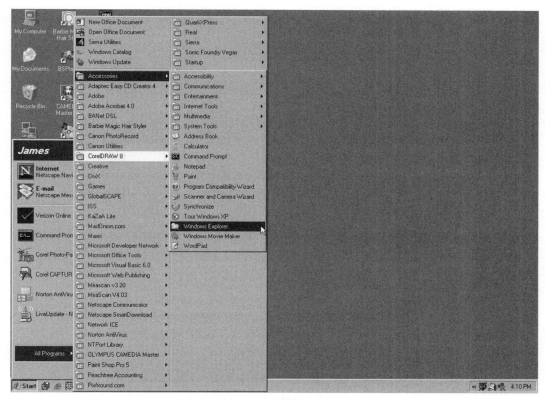

FIGURE 2-2 Finding Windows Explorer in Windows 2000/XP

Place the 3.5-inch PICBasic compiler diskette into your computer's floppy drive, usually the A drive. Now here's where the installation procedure changes. For those using Windows 95/98, start an MS-DOS prompt window. Click Start, select Programs, and then click on MS-DOS Prompt (see Figure 2-8). For Windows ME/2000/XP users, start a command prompt window (equivalent to an MS-DOS prompt window). Then click All Programs, select Accessories, and click Command Prompt (see Figure 2-9).

In either the command prompt window or the MS-DOS window, you will need to type in a few old-fashioned DOS commands. These DOS commands are typed on the command line and the Enter key is hit to execute them.

The DOS instructions are provided to help the reader and serve as supplement to the installation directions provided with the software packages. The instructions are not meant as a DOS tutorial. More information on DOS commands can be found in any

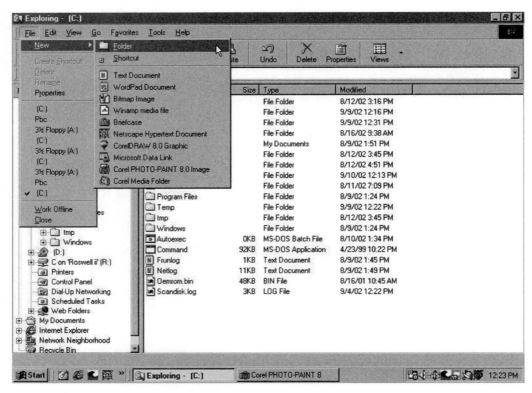

FIGURE 2-3   Creating a new folder (subdirectory) on the C drive

number of DOS manuals. Here is a list of DOS commands we will be using and what action they perform:

Command	Action
cd	Change directory
md	Make directory
copy	Copy files
xcopy	Copy files and subdirectories
path	Sets a search path for executable files
dir	Directory

From this point on, the MS-DOS prompt window and the command prompt window will be referred to as the DOS window. When the DOS window is opened, you will be located in a subdirectory on the hard drive. Your prompt may look like this: C:/WIN-DOWS>.

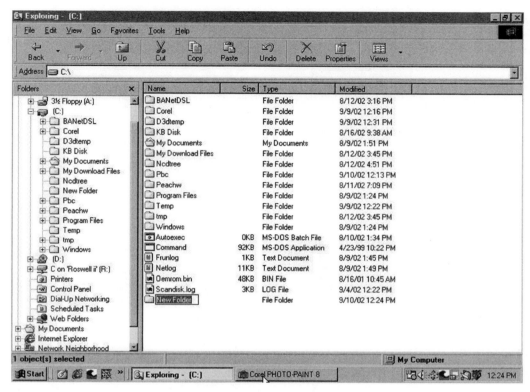

FIGURE 2-4   Type the subdirectory's name in the New Folder icon.

The DOS prompt provides vital information. The C: tells us we are on the C drive. The /WINDOWS tells us we are in the WINDOWS subdirectory.

We want to work from the root directory of the computer's hard drive (usually the C drive). We accomplish this by using the cd (change directory) command.

The cd.. command brings one up a single level in the directory hierarchy. Using cd \ brings one up to the root directory regardless of how deep (as far as levels) one has moved into the subdirectories. The root directory is the top of the directory hierarchy. In other words, from the WINDOWS subdirectory, type in the following command and hit Enter to move to the root directory of the hard drive:

```
cd\
```

We already created our subdirectory PBP using Windows Explorer for the PICBasic Pro compiler. We want to move into the PBP subdirectory, so enter the following command and hit Enter.

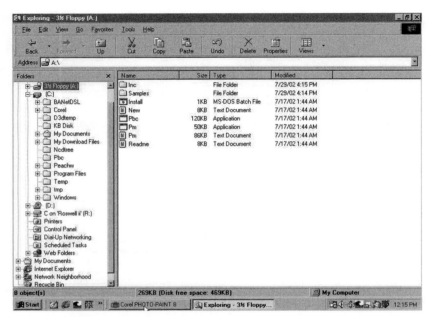

FIGURE 2-5   Selecting the A drive containing the PICBasic program diskette

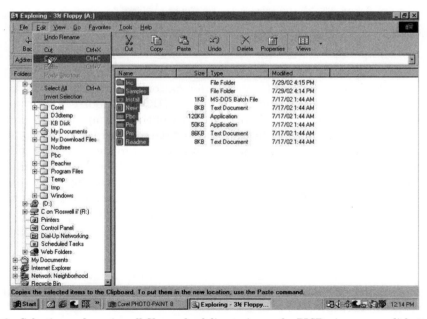

FIGURE 2-6   Selecting and copying all files and subdirectories on the PICBasic program diskette

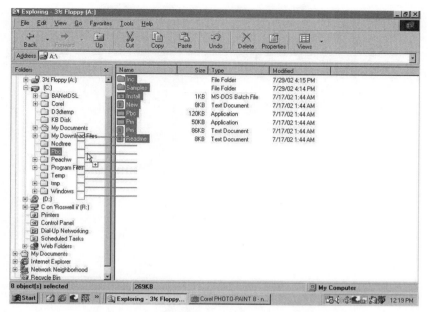

FIGURE 2-7   Copying all the selected files on the PICBasic program diskette in the A drive to the PBC directory on the hard drive

```
C:/> cd pbp
```

Next, place the 3.5-inch PICBasic Pro diskette into your A drive and type the following at the DOS prompt:

```
C:/PCP> A:\pbpxxx -d
```

where xxx is the version number of the compiler on the disk (see Figure 2-10).

This command copies and installs all the required files into the PBP directory. With the files safely loaded onto your hard drive, close the DOS window, remove the diskette, and store it in a safe place in case it is needed in the future.

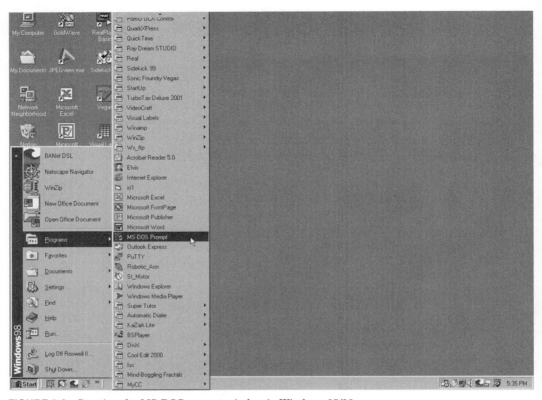

FIGURE 2-8 Starting the MS-DOS prompt window in Windows 95/98

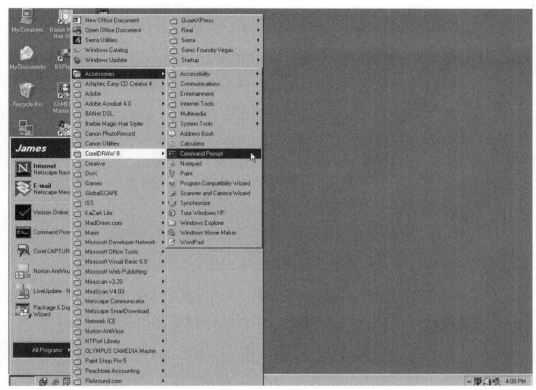

FIGURE 2-9   Starting the Command prompt window in Windows ME/2000/XP

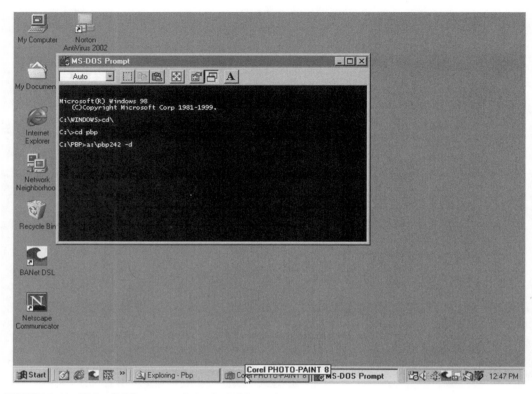

FIGURE 2-10 Using DOS commands in the DOS window prompt to execute the PICBasic Pro installation program

3

Installing the EPIC Software

Installing the EPIC software from Windows is simple. To install it, run the install.bat file on the 3.5-inch EPIC diskette. The install.bat file executes the main, self-extracting program that automatically creates a subdirectory, EPIC, on your computer's hard drive, decompresses the program and its support files, and copies them into the EPIC subdirectory.

If a subdirectory called EPIC already exists on your hard drive when running the install.bat file, you will receive an error message.

If you are still in the same DOS session from the last chapter and want to continue to use DOS to install the EPIC software, skip down to the "Installing EPIC Software from DOS" section. For those who wish to use Windows to install this software, continue reading with the following section.

Installing the EPIC Software in Windows

From Windows, click the Start button and then click Run (see Figure 3-1). Now place the EPIC programming diskette into the A drive, and when the Run menu window opens, select Browse. From the Browse window, select the A drive. This action will list the files on the A drive. Select the install.bat file and click the Open button (see Figure 3-2).

This action brings you back to the Run window, where the install.bat file should be listed (see Figure 3-3). Click OK, which automatically opens a DOS window and starts the executable program on the EPIC diskette. The executable program creates a new subdirectory on the computer's hard drive called EPIC. The program decompresses and copies all the necessary files into the epic subdirectory, as shown in Figure 3-4.

If you installed the EPIC program using Windows, skip over the next section, "Installing the EPIC software from DOS," and continue reading at the "Applications Directory" section.

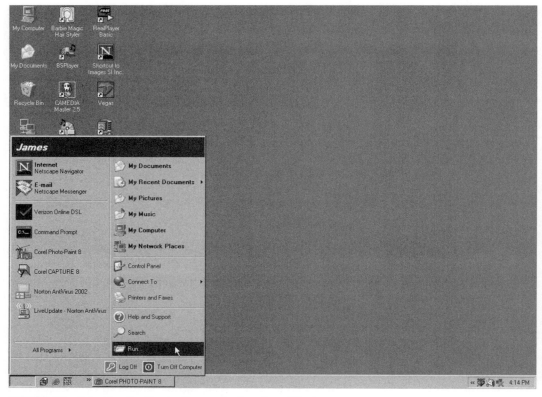

FIGURE 3-1   Selecting the Start > Run menu item from Windows

Installing the EPIC Software from DOS

If you are still operating in the same DOS session from the last chapter, move back into the root directory and enter the following at the prompt:

```
C:/>pbp cd..
```

If you are entering a new DOS window, the prompt may appear a little different, but the command is the same:

```
C:/>WINDOWS cd/
```

From the root directory of the C drive, we will run the install.bat program on the EPIC 3.5-inch diskette. The self-extracting file creates itself a subdirectory called EPIC. Place the EPIC diskette into the floppy drive. At the DOS prompt, enter

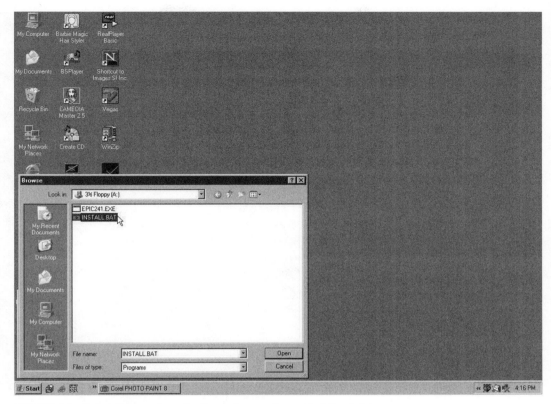

FIGURE 3-2   Select Browse from the Run submenu then select the A drive.

```
C:/> A:
```

This places the command prompt into the A drive. The command prompt should look like this:

```
A:/>
```

Now run the install.bat file by entering the following command:

```
A:/>install.bat
```

This starts the self-extracting file that creates the EPIC subdirectory and installs all the required files into the subdirectory.

With the program and files installed onto your hard drive, remove the diskette and store it in a safe place in case it is needed in the future.

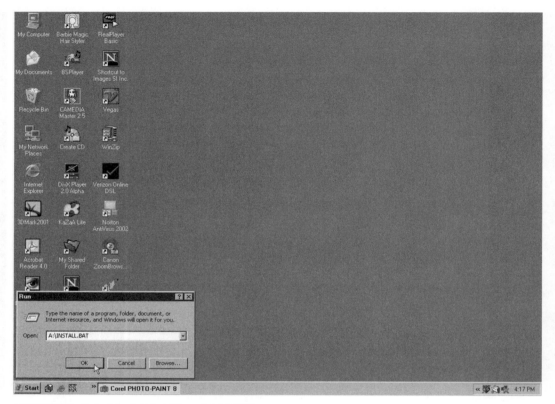

FIGURE 3-3   Select Install.bat file and hit "OK."

Applications Directory

It would be a good idea at this time to create another subdirectory to store all your PICBasic and PICBasic Pro programs. This will keep the PBC (and/or PBP) and EPIC directories clean, neat, and uncluttered with programs and program revisions. From Windows Explorer, create an "Applics" subdirectory on your computer's hard drive.

ZIF Adapter Sockets

After you start programming the samples programs into a 16F84 microcontroller, you will quickly realize that it is inconvenient to continually insert and remove the 16F84 in and out of the standard socket on the EPIC programming board.

However, an 18-pin *Zero Force Insertion* (ZIF) socket adapter for the EPIC board enables you to remove and insert the 16F84 quickly and easily (see Figure 3-5). I recommend purchasing the ZIF adapter because it saves a considerable amount of time and hassle, not to mention bent *integrated circuit* (IC) pins.

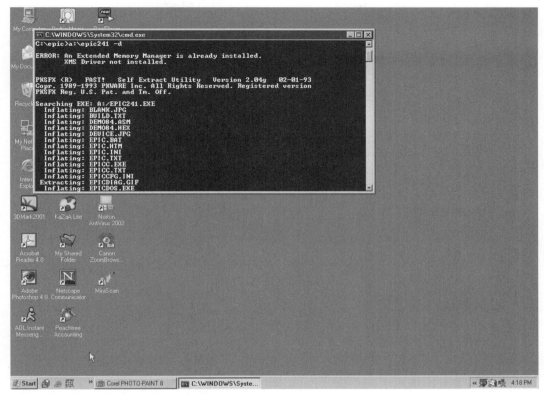

FIGURE 3-4   Install.bat starts a self-extracting program.

AC Adapter

The stock EPIC programming board requires two fresh 9-volt batteries. Although not as critical as the ZIF socket adapter for the EPIC programming board, an AC adapter is available that eliminates the 9-volt batteries. This eliminates any programming problems that will arise as the batteries wear out. These two additions to your programming arsenal will make programming PIC microcontrollers easier.

FIGURE 3-5   ZIF Socket

CodeDesigner

In this chapter we will install, set up, and work with the CodeDesigner software. CodeDesigner is a Windows *integrated development environment* (IDE) interface for the PIC series of microcontrollers. This IDE interface enables you to write code, compile the code, and then program the code into a PIC microcontroller while staying in the same Windows environment.

The compiling of code within CodeDesigner still requires the use of one of the PICBasic compilers, and the compiled code in a PIC microcontroller requires EPIC software and hardware, as does programming. CodeDesigner integrates these software packages and hardware so that they can operate within its Windows environment.

As an editor, CodeDesigner has many useful features that help you write code and that are far superior to simple text editor.

CodeDesigner Features

The features that CodeDesigner offers are as follows:

- **AutoCodeCompletion**   CodeDesigner makes writing code much easier with smart pop-up list boxes that can automatically fill in statements and parameters for you.

- **Multidocument Support**   This includes the Line Error Highlighting option, which compiles your PICBasic Project so that CodeDesigner can read error data and highlight error lines.

- **QuickSyntaxHelp**   The QuickSyntaxHelp feature displays statement syntax when you type in a valid PICBasic statement.

- **Statement Description**   Statement descriptions are displayed in the status bar when you type in a valid PICBasic statement.

- **Statement Help**   Simply position your cursor over a PICBasic statement and get statement-specific help.

- **Label Listbox**   This displays the current label and allows you to select a label from the list to jump to a selected label.
- **Colored PICBasic Syntax**   This sets colors for reserved words, strings, numbers, comments, defines, and so on. The Colored PICBasic Syntax option makes for easy code reading.
- **Bookmarks**   Never lose your place again. CodeDesigner enables you to set bookmarks.
- **Multi-undo/redo**   Didn't want to delete that last line? No problem. Simply click the Undo button.
- **Multiviews**   Multiple views of your source code enable you to easily edit your code.
- **Print Source Code**
- **Drag and Drop Text**
- **Row/column-based Insert, Delete, and Copy**
- **Search and Replace**
- **Compile and launch device programmer**

One feature I like is that each typed line of code is color-coded, making it easier to spot errors and read through your code.

When you purchase either the PICBasic or PICBasic Pro compilers, they are packaged with an additional diskette that contains a freebie version of CodeDesigner called CodeDesigner Lite. The Lite version enables you to write programs up to 150 lines and open up 3 source files at once for easy copy and paste. If you would like to try CodeDesigner without purchasing a compiler, CodeDesigner Lite is freely downloadable from www.imagesco.com/catalog/pic/codedesigner.html.

The idea is that, if you like the freebie CodeDesigner software, you can then upgrade to the full-featured CodeDesigner. The standard version of CodeDesigner costs $75 and removes the restrictions imposed in the Lite version. This standard version enables you to write programs with an unlimited amount of code lines and to open an unlimited amount of source files—unlimited with respect to the limits of your computer capabilities.

If, for any reason, you do not wish to use the CodeDesigner software, the procedures for writing code, compiling, and programming a PICmicro chip manually from a DOS environment are covered in Chapter 5, "How to Use DOS Instead of Windows to Code, Compile, and Program."

CodeDesigner increases productivity and the ease with which to write, debug, and load PICBasic programs into the microcontroller. If a problem occurs (more often than not), debugging the code and recompiling are much easier and faster using CodeDesigner. When the program is completely debugged, it can be uploaded into the PIC microcontroller using the EPIC software and programming board. At this point, the microcontroller and circuit are tested. If it functions properly, you're finished. If not, you must begin rewriting the program or redesigning the electronics.

Software Installation

The CodeDesigner software loads like most standard Windows software. Load the software on your computer's hard drive according to the instructions.

When CodeDesigner installs, it creates a subdirectory in the Program Files directory. It places a CodeDesigner shortcut on the Start and Program menus in Windows.

Setting CodeDesigner Options

In order for CodeDesigner to compile code and program the resulting code into PIC microcontrollers, you need to configure the default locations where CodeDesigner looks for its support programs. You can set up the default locations by entering the software paths where CodeDesigner stores programs, where it looks for PICBasic compiler, and where the EPIC program can be found.

Start the CodeDesigner software and the evaluation copy opens with the version window shown in Figure 4-1. The next window is the standard opening screen to the

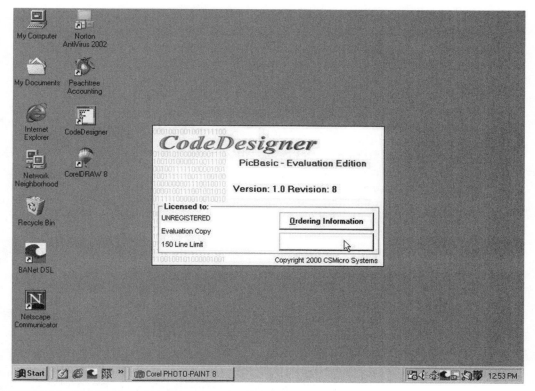

FIGURE 4-1   CodeDesigner Lite startup version screen

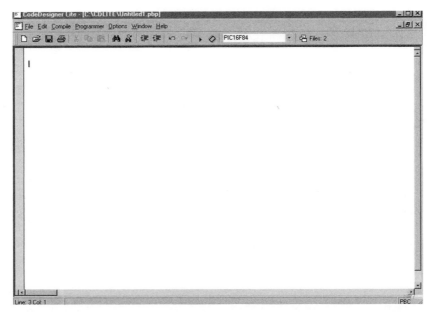

FIGURE 4-2   Opening screen in CodeDesigner ready for writing, compiling, and programming

CodeDesigner software (see Figure 4-2). To begin setting the options, click the Compile menu option and then Compiler Options (see Figure 4-3).

The Compiler Options window opens, as shown in Figure 4-4. In the top text field, use the pull-down menu to choose which compiler you are using, the PICBasic Pro or PICBasic. In Figure 4-4, the PICBasic Pro compiler is chosen.

In the second text field, you select the compiler pathname. The compiler path and name (PBPW.exe) are chosen for the PICBasic Pro compiler in the subdirectory of C:\PBP.

In the third text field, choose where the CodeDesigner software will load and save your source code files. Hit the Browse button next to the text field. This opens a browser window, as shown in Figure 4-5. Select the Applics subdirectory on the hard drive and click OK.

The Default Source Code Directory text field now contains the path C:\Applics subdirectory (see Figure 4-6). Click OK to close the Compiler Options window.

Now you need to set the Programmer options. Click on the Programmer menu and select Programmer Options (see Figure 4-7), which opens the Programmer Options window (see Figure 4-8). Click the Browse button next to the Programmer Pathname text field. This opens a browser window where you will select the EPICWIN program in the EPIC subdirectory on your computer's hard drive (see Figure 4-9). Click Open, which

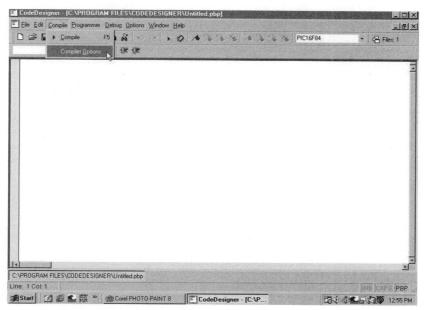

FIGURE 4-3   CodeDesigner Compiler Options menu item

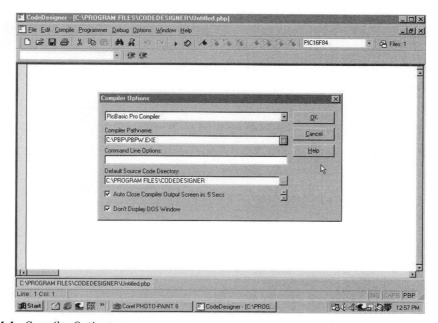

FIGURE 4-4   Compiler Options menu

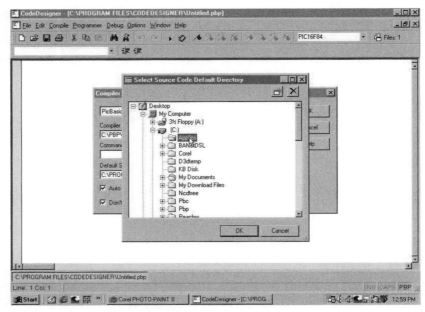

FIGURE 4-5 Selecting the Applics subdirectory for the source code files

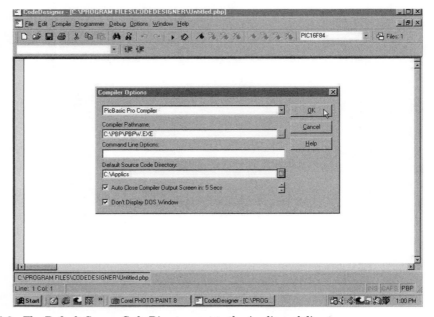

FIGURE 4-6 The Default Source Code Directory set to the Applics subdirectory

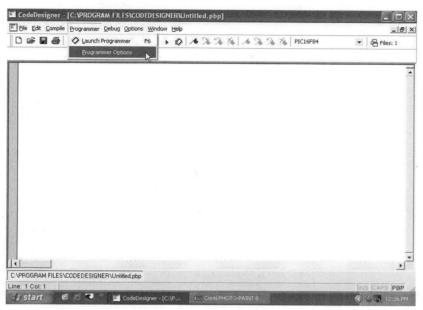

FIGURE 4-7 CodeDesigner Programmer Option menu item

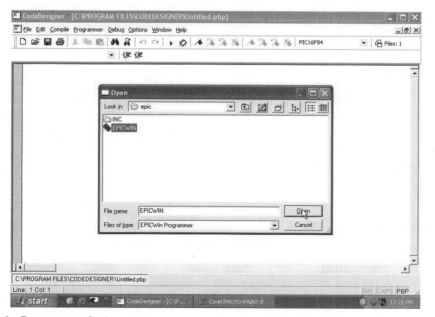

FIGURE 4-8 Programmer Options menu

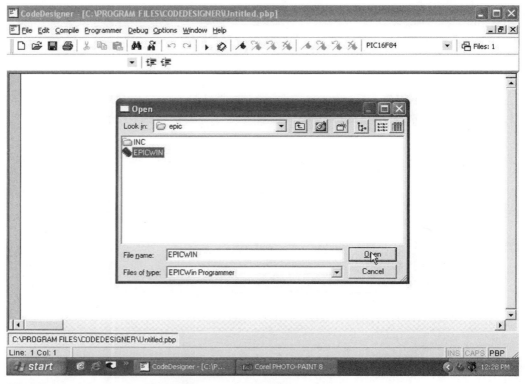

FIGURE 4-9   Selecting the EPICWIN program in the EPIC subdirectory

brings you back to the Programmer Options window. The new path you select should be in the Programmer Pathname text field (see Figure 4-10). Click OK to set this option.

With the CodeDesigner options set, you are ready to write your first program.

First Program

Start CodeDesigner and enter the following code for the PICBasic compiler:

```
' PICBasic program
' Wink Program
' Blinks and winks two LEDs connected to PORT B.
Loop:
High 0        ' Turn on LED connected to pin RB0
Low 1         ' Turn off LED connected to pin RB1
Pause 500     ' Delay for .5 seconds
Low 0         ' Turn off LED connected to pin RB0
High 1        ' Turn on LED connected to pin RB1
```

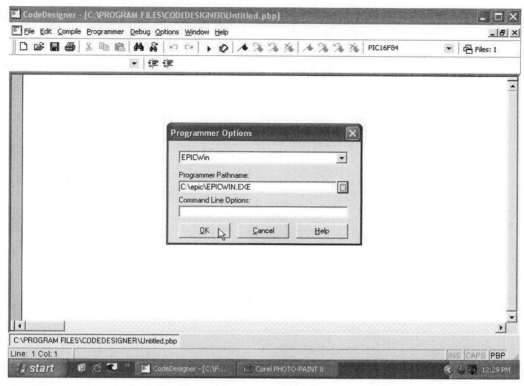

FIGURE 4-10 Programmer pathname set to EPICWIN.exe program

```
Pause 500      ' Delay for .5 seconds
GoTo Loop      ' Go back to loop and blink & wink LEDs forever
End
```

The next program is identical in function (not code) to the previous PICBasic program. Start CodeDesigner and enter the following code (see Figure 4-11):

```
' PICBasic Pro program
' Wink Program
' Blinks and winks two LEDs connected to PORT B.
Loop:
High PORTB.0   ' Turn on LED connected to RB0
Low PORTB.1    ' Turn off LED connected to RB1
Pause 500      ' Wait 1/2 second
Low PORTB.0    ' Turn off LED connected to RB0
High PORTB.1   ' Turn on LED connected to RB1
Pause 500      ' Wait 1/2 second
GoTo Loop      ' Loop back- repeat cycle blink & wink forever
```

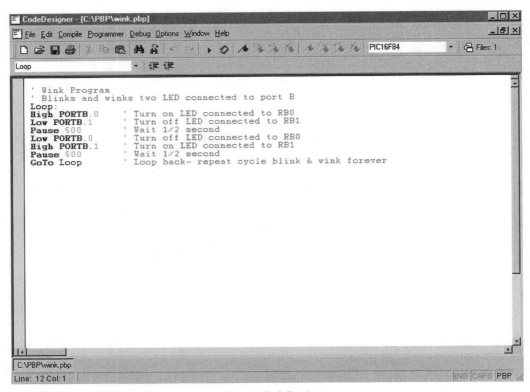

FIGURE 4-11   A PICBasic Pro program written in CodeDesigner

NOTE: We could have used the PICBasic program for the PICBasic Pro compiler also. The Pro compiler would understand the High X and Low X commands and pin numbers, which will be explained later. In the meantime it is better programming practice for the PICBasic Pro compiler to clearly define the port pins as used in the Pro-style program.

CodeDesigner defaults to writing code for the PIC 16F84 microcontroller, which is the microcontroller I recommend to start with. To change the microcontroller, simply pull down the device menu and select the appropriate microcontroller (see Figure 4-12).

When CodeDesigner attempts to compile a program from the Windows environment, it automatically opens a DOS prompt window, compiles the program, and then ends the DOS session.

To compile the program using CodeDesigner, either select Compile under the Compile menu or hit F5. CodeDesigner automatically starts the PICBasic Pro compiler (or

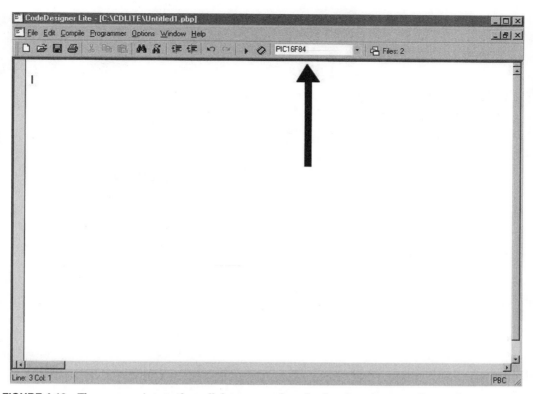

FIGURE 4-12   The arrow points to the pull-down menu for selecting the microcontroller (16F84).

PICBasic compiler) to compile the program. Before you attempt to compile a program, make sure you have set up the "compiler options" under the Compile menu.

Once the program is compiled, you can go to the next step of loading the program into a PIC microcontroller chip.

First, connect the EPIC programming board to the printer port. If your computer has only one printer port, disconnect the printer, if one is connected, and attach the EPIC programming board using a six-foot DB25 cable.

When connecting the EPIC programming board to the computer, there should not be any microcontroller installed on the board. If you have an AC adapter for the EPIC programmer board, plug it into the board. If not, attach two fresh 9-volt batteries. Connect the "Batt ON" jumper to apply power. The programming board must be connected to the printer port with power applied to the programming board before starting the EPIC programming software. If not, the software will not know that the programming board is connected to the printer port and give the error message "EPIC Programmer not found" (see Figure 4-13).

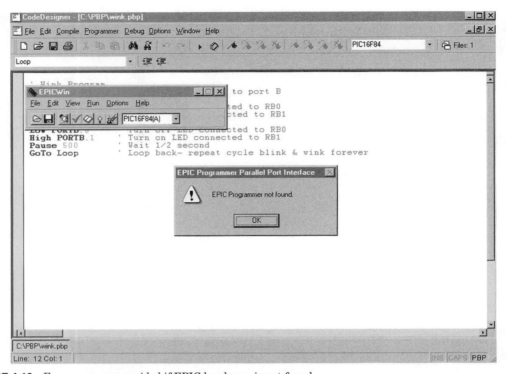

FIGURE 4-13   Error message provided if EPIC hardware is not found

The EPIC Programming Board Software

To program the 16F84 microcontroller from within CodeDesigner, select the Launch Programmer menu item from the Programmer menu (found in the CodeDesigner taskbar) or hit F6. CodeDesigner automatically starts the EPICWIN.exe Windows software.

With the EPIC windows software started, set the configuration switches one by one under the Configuration menu (see Figure 4-14):

- **Device**   Sets the device type. Set it for 16F84 (default).
- **Memory Size (K)**   Sets the memory size. Choose 1.
- **OSC**   Sets the oscillator type. Choose XT for crystal.
- **Watchdog Timer**   Choose On.
- **Code Protect**   Choose Off.
- **Power Up Timer Enable**   Choose High.

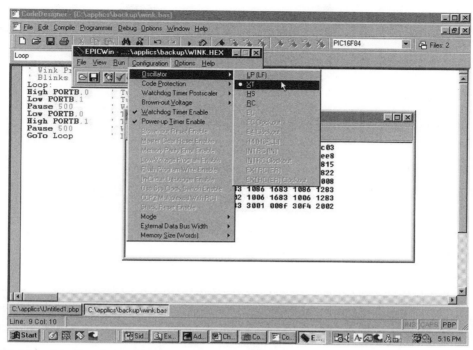

FIGURE 4-14   Selecting the configuration switches in EPICWIN

After the configuration switches are set, insert the PIC 16F84 microcontroller into the open socket on the EPIC programming board.

If you receive an error message "EPIC Programmer not found" when CodeDesigner starts the EPIC windows program (refer to Figure 4-13), you have the option of either troubleshooting the problem or using the EPIC DOS program. For instructions on using EPIC software (the DOS version), see Chapter 5.

The schematic of the circuit needed to test the PICmicro is given in Chapter 6, "Testing the PIC Microcontroller." If you have successfully written, compiled, and uploaded the code into the PICmicro chip using CodeDesigner, you can skip the DOS material in Chapter 5 and go straight to Chapter 6.

Parts List

CodeDesigner Lite   A free download is available from www.imagesco.com/catalog/pic/codedesigner.html.

Code Designer Standard   Available for $75 from Images SI, Inc., 109 Woods of Arden Road, Staten Island, NY 10312, 718-698-8305, 718-966-3695 fax.

5

How to Use DOS Instead of Windows to Code, Compile, and Program

In Chapter 4, "CodeDesigner," you compiled and programmed your microcontroller using the CodeDesigner program. If, for some reason, you do not wish to use or cannot use CodeDesigner Lite, this chapter will instruct you in how to perform all the functions for writing, compiling, and programming the code in a PICmicro chip from DOS or a DOS prompt window.

When starting a new DOS session, use the Path command (see Figure 5-1), so you will not have to copy and swap files back and forth across directories. If you have created the directory names as suggested, you can use the following command.

For PICBasic users, the command is

```
path \;c:\pbc;c:\epic;c:\windows\command;
```

For PICBasic Pro users, the command is

```
path \;c:\pbp;c:\epic;c:\windows\command;
```

Now we can begin by using a standard word processor or text editor to write the PICBasic program in DOS. Windows users can use the Notepad program, and DOS users can use the Edit program. In DOS, we will work from and store our program(s) in the subdirectory we created earlier called Applics.

Move into the Applics subdirectory. Enter the cd (change directory) command at the DOS prompt (refer to Figure 5-1):

```
C:\> cd applics
```

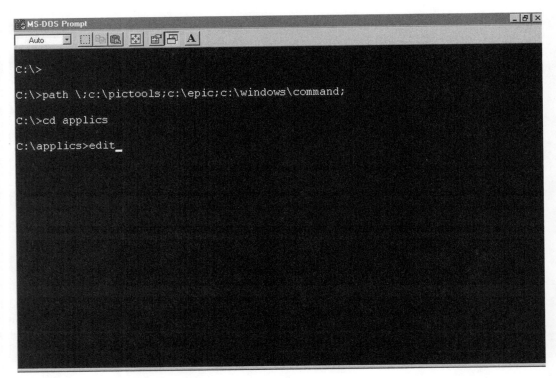

FIGURE 5-1   Entering DOS commands for paths, changing directories, and starting the Edit program

Once in this directory, the prompt changes to

```
C:\applics>
```

In this example, I will be using the free EDIT program packaged with Windows to write the program. Start Edit by typing "edit" at the command prompt (refer to Figure 5-1):

```
C:\applics> edit
```

This starts the Edit program, as shown in Figure 5-2.
Enter this program in your word processor exactly as it is written:

```
'1st PICBasic program
'Wink two LEDs connected to PORT B.
Loop:High 0     ' Turn on LED connected to pin RB0
Low 1           ' Turn off LED connected to pin RB1
Pause 500       ' Delay for .5 seconds
Low 0           ' Turn off LED connected to pin RB0
High 1          ' Turn on LED connected to pin RB1
```

FIGURE 5-2 Start screen of Edit program

```
Pause 500        ' Delay for .5 seconds
Goto loop        ' Go back to loop and blink & wink LEDs forever
End
```

Save the text file in the Applics directory, as shown in Figure 5-3. Use the Save function under the File menu and name the file wink.bas (see Figure 5-4). If, by accident, you save the file as wink.txt, don't get discouraged. You can do a Save As from the Edit program (under the File menu) and rename the file wink.bas.

For PICBasic Pro users, enter the following text in your word processor and save the file as wink.bas:

```
' 1st PICBasic Pro Program
' Winks two LEDs connected to port B
Loop:
High PORTB.0    ' Turn on LED connected to RB0
Low PORTB.1     ' Turn off LED connected to RB1
Pause 500       ' Wait 1/2 second
Low PORTB.0     ' Turn off LED connected to RB0
High PORTB.1    ' Turn on LED connected to RB1
Pause 500       ' Wait 1/2 second
GoTo Loop       ' Loop back- repeat cycle blink & wink forever
```

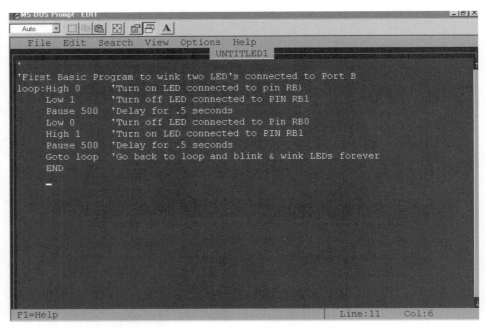

FIGURE 5-3   Entering the wink.bas program

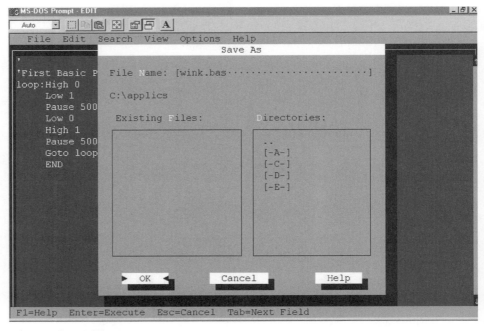

FIGURE 5-4   Saving the wink.bas program

Compile

The PICBasic compiler (or PICBasic Pro compiler) can be run from DOS or from a DOS prompt window within Windows, but here you will run the PICBasic compiler from the Applics directory. Type the command "pbc -p16f84 wink.bas" at the DOS prompt and hit Enter (see Figure 5-5):

```
C:/APPLICS>pbc -p16F84 wink.bas
```

For PICBasic Pro, the command is "C:/APPLICS>pbp -p16F84 wink.bas."

The compiler displays an initialization copyright message and begins processing the Basic source code (see Figure 5-6). If the Basic source code is without errors (and why shouldn't it be?), it will create two additional files. If the compiler finds any errors, a list of errors with their line numbers will be displayed. Use the line numbers in the error message to locate the line numbers in the .bas text file where the errors occurred. The errors need to be corrected before the compiler can compile the source code correctly. The most common errors are with Basic language syntax and usage.

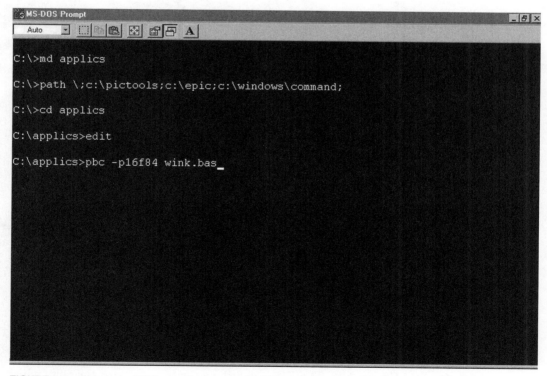

FIGURE 5-5   Entering a DOS command to run the PICBasic compiler program on the wink.bas program for the 16F84 microcontroller

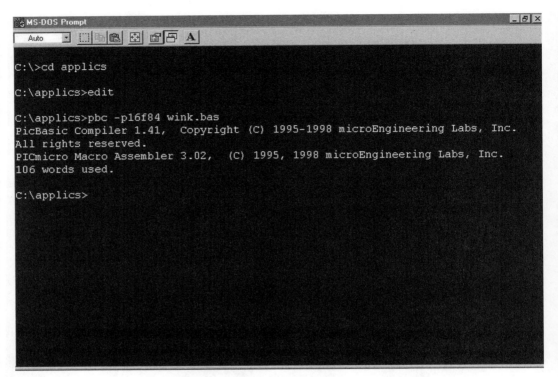

FIGURE 5-6   Typical copyright notice and notice provided by the PICBasic compiler when it is run successfully

You can look at the files by using the dir directory command. Type dir at the command prompt

```
C:\APPLICS> dir
```

and hit Enter (see Figure 5-7).

The dir command displays all the files and subdirectories within the subdirectory where it is issued. In Figure 5-7, the two additional files the compiler created are displayed. One file is the wink.asm file and is the assembler source code file that automatically initiated the macro-assembler to compile the assembly code to machine language hex code. The hex code file is the second file called wink.hex.

Programming the PIC Microcontroller Chip

To program the PIC chip, we must connect the EPIC programming carrier board to the computer (see Figure 5-8). The EPIC board connects to the printer port, which is also

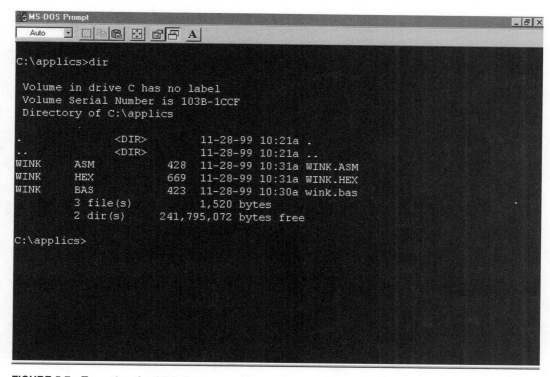

```
C:\applics>dir

 Volume in drive C has no label
 Volume Serial Number is 103B-1CCF
 Directory of C:\applics

.              <DIR>        11-28-99 10:21a .
..             <DIR>        11-28-99 10:21a ..
WINK    ASM         428     11-28-99 10:31a WINK.ASM
WINK    HEX         669     11-28-99 10:31a WINK.HEX
WINK    BAS         423     11-28-99 10:30a wink.bas
        3 file(s)         1,520 bytes
        2 dir(s)     241,795,072 bytes free

C:\applics>
```

FIGURE 5-7 Executing the DOS dir command to see the two additional files (.ASM and .HEX) created by the PICBasic compiler

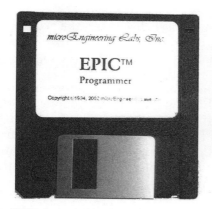

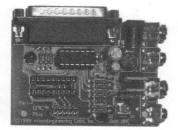

FIGURE 5-8 EPIC programming board and software

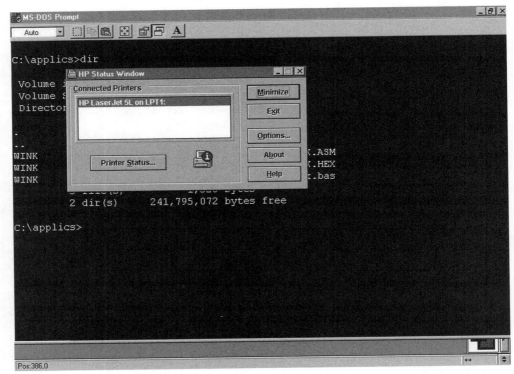

FIGURE 5-9 Message box window used to temporarily close the printer driver to provide better access to the printer port for the EPIC programmer

called the parallel port, and either name may be used. A computer may contain up to four parallel (printer) ports. Each port is assigned a number one through four. The computer lists these ports as LPT1 to LPT4.

If your computer has only one printer port, disconnect the printer, if one is connected, and attach the EPIC programming board using a six-foot DB25 cable. In some cases, it may be necessary to temporarily remove the printer driver. Figure 5-9 shows a typical window to disable an HP printer.

When connecting the programming board to the computer, make sure no PIC microcontroller is installed on the board. If you have an AC adapter for the EPIC programmer board, plug it into the board. If not, attach two fresh 9-volt batteries. Connect the "Batt On" jumper to apply power. The programming board must be connected to the printer port with power applied to the programming board before starting the software. If not, the software will not see the programming board connected to the printer port and give the error message "EPIC Programmer not connected."

When power is applied and it is connected to the printer port, the *light-emitting diode* (LED) on the EPIC programmer board may be on or off. Do not insert a PIC microcontroller into the programming board socket until the EPIC programming software is running.

The EPIC Programming Board Software

Two versions of the EPIC software are available: EPIC.exe for DOS and EPICWIN.exe for Windows. The Windows software is 32 bit. It may be used with Windows 95, Windows 98, and Windows NT/XP, *but not* Windows 3.X. It has been my experience that Windows 95/98 printer drivers many times like to retain control of the printer (LPT1) port. If this is the case with your computer, the Windows EPIC program may not function properly and you may be forced to use the DOS-level program. If you receive the error message "EPIC Programmer not connected" when you start the EPIC windows program, you have the option of either troubleshooting the problem or using the EPIC DOS program.

Using the EPIC DOS Version

If using Windows 95 or higher, you could either open a MS-DOS prompt window or restart the computer in the DOS mode. Windows 3.X users should end the Windows session.

Continuing with the WINK.BAS Program

Assuming you are still in the same DOS session and have just run the PBC compiler on the WINK.BAS program, you are still in the applics directory. At the DOS prompt, type "epic" and hit Enter to run the DOS version of the EPIC software (see Figure 5-10).

If you are operating out of a DOS window, you may get a Device Conflict message box, as shown in Figure 5-11. You want MS-DOS to control the LPT port so the EPIC programming software will work. Select the MS-DOS Prompt and hit OK.

EPIC's opening screen is shown in Figure 5-12. Click the Open File button or press Alt-O on your keyboard and select the wink.hex file (see Figure 5-13). When the hex file loads, you will see a list of numbers in the window on the left (see Figure 5-14). This is the machine code of your program. On the right-hand side of the screen are configuration switches that you need to check before you program the PIC chip.

Let's go through the configuration switches once more:

- **Device**   Sets the device type. Set it for 8X.
- **ROM Size (K)**   Sets the memory size. Choose 1.
- **OSC**   Sets the oscillator type. Choose XT for crystal.
- **Watchdog Timer**   Choose On.
- **Code Protect**   Choose Off.
- **Power Up Timer Enable**   Choose High.

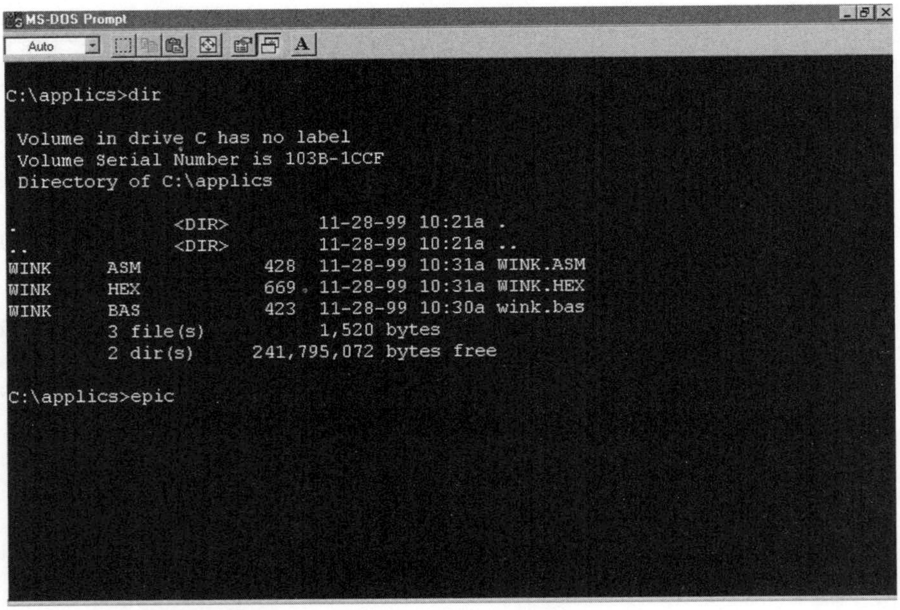

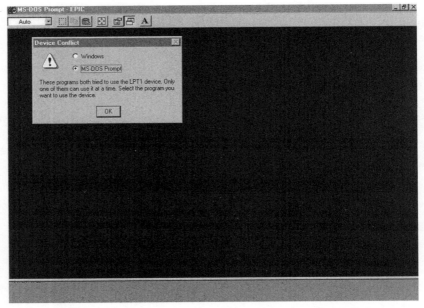

```
MS-DOS Prompt
 Auto

C:\applics>dir

 Volume in drive C has no label
 Volume Serial Number is 103B-1CCF
 Directory of C:\applics

.              <DIR>         11-28-99 10:21a .
..             <DIR>         11-28-99 10:21a ..
WINK     ASM        428      11-28-99 10:31a WINK.ASM
WINK     HEX        669      11-28-99 10:31a WINK.HEX
WINK     BAS        423      11-28-99 10:30a wink.bas
         3 file(s)            1,520 bytes
         2 dir(s)       241,795,072 bytes free

C:\applics>epic
```

FIGURE 5-10   Entering the DOS epic command to start the program

```
MS-DOS Prompt - EPIC
 Auto

 Device Conflict
  ⚠    ○ Windows
       ● MS-DOS Prompt

  These programs both tried to use the LPT1 device. Only
  one of them can use it at a time. Select the program you
  want to use the device.

              [ OK ]
```

FIGURE 5-11   Possible Device Conflict error message when DOS and Windows both try to use the printer port. Select DOS and OK.

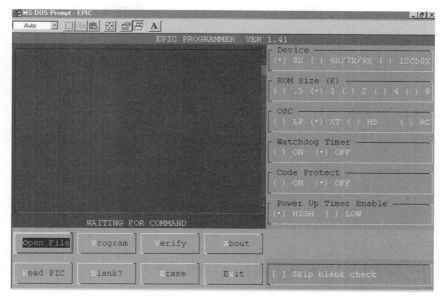

FIGURE 5-12   Opening screen of the EPIC programming software. Hit the Open File button.

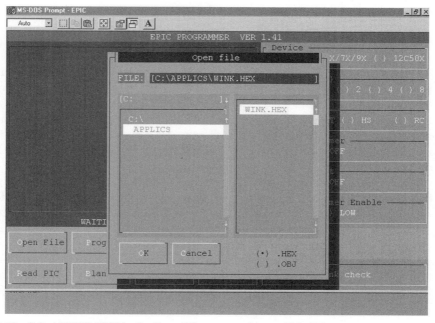

FIGURE 5-13   Select WINK.HEX in the Open File message box and hit OK.

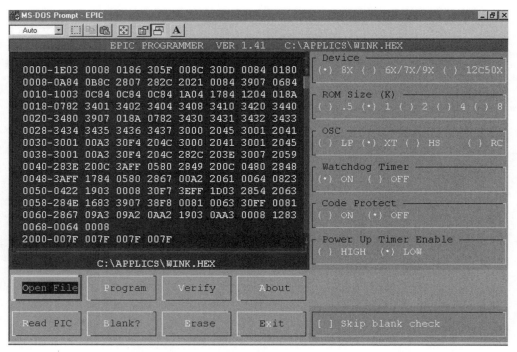

FIGURE 5-14   The hexadecimal numbers in the EPIC window are the machine-language version of the wink.bas program that is uploaded (programmed) into the 16F84 microcontroller.

After the configuration switches are set, insert the PIC 16F84 microcontroller into the socket. Click Program or press Alt-P on the keyboard to begin programming. The EPIC program first looks at the microcontroller chip to see if it is blank. If it is, the EPIC program installs your program into the microcontroller. If the microcontroller is not blank, you are given the option to cancel the operation or overwrite the existing program with the new program. If an existing program is contained in the PIC chip's memory, write over it.

I have noticed that when I place a brand-new PICmicro 16F84 chip into the EPIC compiler to program, EPIC always reports existing code on the chip. I don't know if this is a bug in the EPIC software or if Microchip Technologies loads numbers into the chip's memory for testing purposes. Don't let it throw you—the PICmicro chip is new.

The machine language code lines are highlighted as the EPIC software uploads the program into the PIC microchip. When it is finished, the microcontroller is programmed and ready to run. You can verify the program if you like by hitting (or highlighting) the Verify button. This initiates a comparison of the program held in memory to the program stored in the PIC microcontroller.

Chapter 6

Testing the PIC Microcontroller

In this chapter we will build and test the circuit for the PIC microcontroller we programmed. The components needed to build the circuit were listed way back in Chapter 1, "Microcontrollers." If you purchased the components, you can quickly set up the test circuit and run the program. If not, the components are relisted at the end of this chapter.

The Solderless Breadboard

A solderless breadboard is a useful component for building test circuits and prototyping. For those of us who have not dabbled in electronics very much, the solderless breadboard will be described in detail (see Figure 6-1).

As its name implies, you can breadboard (assemble and connect) electronic components onto the breadboard without the use of soldering. Once you build your circuit, you can change, modify, or remove components from the circuit at any time. This makes it easy to correct any design or wiring errors. The solderless breadboard is reusable. Once a circuit is tested and works to expectations, the circuit may be removed from the solderless breadboard, hardwired, and soldered onto a *printed circuit board* (PCB). You will find a solderless breadboard a valuable asset for building and testing the circuits outlined in this book.

In Figure 6-2, a partial cutaway of the top surface shows some of the internal structure of the breadboard. The holes on the board are plug-ins for wires. When a wire or pin is inserted into a hole, it makes intimate (electrical) contact with the metal connector strip inside the breadboard. The holes are properly spaced (.1 inch × .1 inch) so many integrate circuits and other components can be plugged in. You connect components to one another on the board using 22-guage (solid or stranded) wire. I prefer to use stranded wire because it has greater flexibility, whereas other people prefer solid wire because it's stiffer and easier to push into breadboard holes.

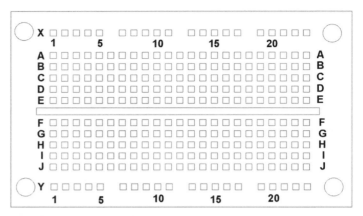

FIGURE 6-1   Top view of a solderless breadboard

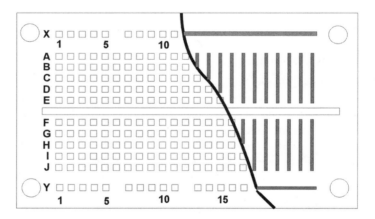

FIGURE 6-2   Top view of a solderless breadboard with a partial cutaway showing the conductive contact strips underneath

The complete internal wiring structure of the solderless boards is shown in Figure 6-3. The left side of the solderless breadboard shows the X and Y rows that are typically used to supply power (*voltage at the common collector* [Vcc]) and ground connections to the circuit. The columns below the X row and above the Y row are used for mounting components. The right side of the solderless breadboard shows double rows located at the top and bottom of the board. This is used to supply both Vcc and ground on each side of the breadboard.

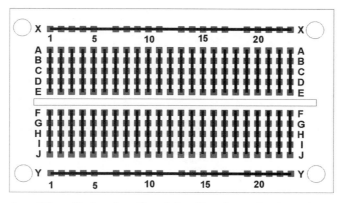

FIGURE 6-3   Top view of the solderless breadboard detailing the conductive strips

Three Schematics, One Circuit

Figures 6-4, 6-5, and 6-6 are identical schematics of our test circuit. The 16F84 PIC microcontroller in the schematic is the microcontroller you programmed in either Chapter 4, "CodeDesigner," or Chapter 5, "How to Use DOS Instead of Windows to Code, Compile, and Program." I drew three schematics to help orient experimenters who may not be familiar with standard electrical drawings. Figure 6-4 shows how the PIC 16F84 microcontroller and components appear. A legend at the bottom shows the electrical symbol and the typical appearance of the component. Figure 6-5 is a line drawing showing how the components appear mounted on one of the solderless breadboards, and each electrical component is pointed out.

If you examine the placement of the components mounted on the solderless breadboard with its internal electrical wiring (refer to Figures 6-2 and 6-3), you can see how the components connect to one another and produce the electrical circuit.

Figure 6-6 is the same schematic drawn as a standard electrical drawing with the pin numbers grouped and orientated to function. For the remainder of the book, standard electrical drawings will be used. Figure 6-6 shows how minimal the components are to get your microcontroller up and running. Primarily, you need a pull-up resistor on PIN 4 (MCLR), a 4 MHz crystal with two (22 pF) capacitors, and a 5-volt power supply.

NOTE: The 4 MHz crystal and two (22 pF) capacitors make up an oscillator that is required by the microcontroller. These three parts may be substituted with a 4 MHz ceramic resonator.

The two *light-emitting diodes* (LEDs) and the two resistors connected in a series with each LED are the output. They enable you to see that the microcontroller and program are functioning properly.

Assemble the components as shown in Figure 6-5 and then move on to the solderless breadboard. When you are finished, your work should appear as in Figure 6-7.

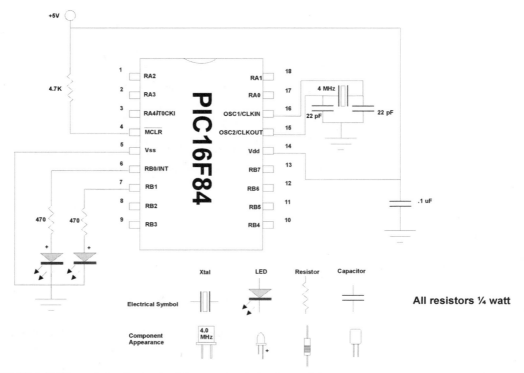

FIGURE 6-4   Isometric schematic of the test circuit for the wink.bas program

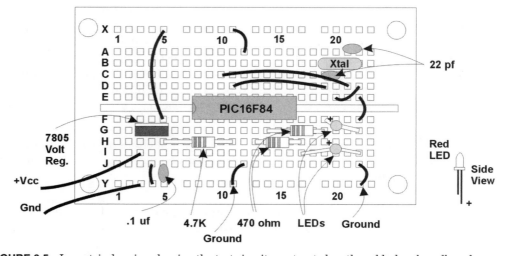

FIGURE 6-5   Isometric drawing showing the test circuit constructed on the solderless breadboard

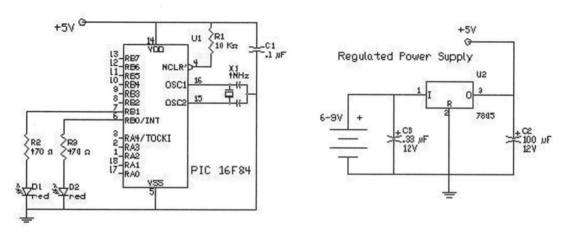

FIGURE 6-6 Schematic of the test circuit for the wink.bas program

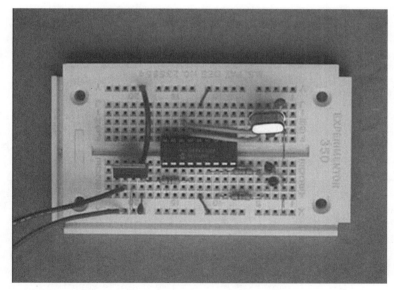

FIGURE 6-7 Wink.bas circuit constructed on a solderless breadboard

Although the specifications sheet on the 16F84 states the microcontroller will operate on voltages from 2 to 6 volts, I provided a regulated 5-volt power supply for the circuit. The regulated power supply consists of a 7805 voltage regulator and two filter capacitors.

Wink

Now apply power to the circuit. The LEDs connected to the chip will alternately turn on and off. Wink, wink . . . now you know how easy it is to program these microcontrollers and get them up and running.

Troubleshooting the Circuit

Not too much can go wrong here. If the LEDs do not light, the first thing to check is their orientation. If they are put in backwards, they will not light.

Next, check your ground wires. Examine the jumper wires on the right-hand side of the solderless breadboard. They bring the ground up to the two (22 pF) capacitors.

Check all your connections. Look back at Figures 6-2 and 6-3 to see how the underlying conductive strips relate to the push-in sockets on top of the board.

As stated, if you wish to make any circuit permanent, you can transfer the components onto a PCB and solder the components together with the foreknowledge that the circuit functions properly. To this end, RadioShack sells a standard PCB (part number 270-170). The copper traces on the underside of the Radio Shack PC board are lined up in a similar manner to the electrical connections on the solderless breadboard. This makes it easier to transfer circuits from the solderless breadboard to the PC board.

PIC Experimenter's Board and LCD

You may want to own two optional tools if you plan on experimenting with the PIC 16F84 and microcontrollers in general. These are the PIC experimenter's board and the *liquid crystal display* (LCD). The LCD serial module will be examined first because a similar LCD is incorporated in the experimenter's board and everything said about the serial LCD module is also true for the PIC experimenter's board LCD. Chapter 14, "Reading Resistive Sensors," will look at LCD modules again and explain LCD commands that are compatible to both the serial LCD modules and the PIC experimenter's board LCD.

PIC microcontrollers lack a display. With a display, the microcontroller can output textual and numeric messages to the user to show how a program is running or what it is detecting.

Serial LCDs on the market only require a single serial line and a common ground. The particular LCD we are using receives standard serial data (RS-232) at 300, 1,200, 2,400, and 9,600 baud (inverted or true). The LCD module is a 2-line by 16-character visible display. The full display is actually 2 lines by 40 characters, but the additional 24 characters per line are offscreen. We can use the PICBasic and PICBasic Pro Serout command to communicate and output messages to the LCD.

The PICBasic and PICBasic Pro compilers can send and receive serial information at 300, 1,200, 2,400, and 9,600 baud. Data is sent as 8 bits, no parity, and 1 stop bit. The serial mode may be set to be true or inverted. This data matches the serial communication protocols required of the LCD.

The LCD module has three wires: +5-volt (red), ground (GND, black or brown), and a serial in line (white). The baud rate may be set to 300, 1,200, 2,400, or 9,600 (true or inverted) using a set of jumpers (J1, J2, and J3) on the back of the LCD.

The LCD module must match the baud rate and mode of the serial data it will receive from the computer/microcontroller in order for the information to be displayed. The LCD module has a self-testing mode that will print the current baud rate as determined by the jumper settings and mode (true/inverted). To enter self-test mode, connect the serial in line to ground (for true) or +5-volt (for inverted) upon LCD module startup.

This first program prints the message "Hello World." The cursor (printing position) automatically moves from left to right. The schematic is shown in Figure 6-8 and the LCD is shown in Figure 6-9. The baud rate is set at 1,200 baud true.

```
'PICBasic Program 6.1 --LCD Test --
Pause 1000                               'Wait 1 second for LCD to initialize
start:
Serout 1, T1200, (254,1)                 'Clear screen
Pause 40
Serout 1, T1200, ("Hello World")         'Print message
pause 400
goto start
End
```

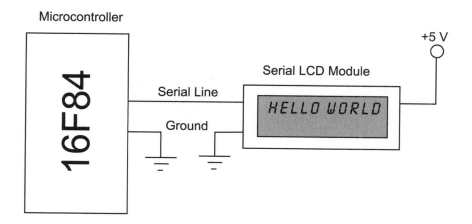

FIGURE 6-8 Schematic of LCD test circuit

FIGURE 6-9 LCD of "Hello World"

This program was kept small to show how easy it is to print a message on the LCD. The following is the same program written for the PICBasic Pro compiler (1,200 baud true):

```
PICBasic Pro Program 6.2 -- LCD Test --
Pause 1000                              'Wait 1 second for LCD to initialize
start:
Serout Portb.1, 1, [254,1]              'Clear Screen
Pause 40
Serout Portb.1, 1, ["Hello World"]      'Print Message
pause 400
goto start
end
```

Notice that line 4 of the program "Serout 1, T1200, (254,1)" is an LCD command. LCD commands are explained further in Chapter 14.

If you print past the 16 displayed characters, the following 24 characters will not show on the LCD screen, but they are in the LCD in the offscreen memory area. The alphanumeric LCD modules you are using have 80 bytes of memory, arranged appropriately for a 2×40 screen. You can use the cursor-positioning instructions (see Chapter 14) to print to a particular location on the display, or you can deliberately print in offscreen memory to temporarily hide text and then send scroll-left instructions to reveal it (see Chapter 14).

PIC Experimenter's Board

The PIC experimenter's board is a prefabricated developing board for prototyping circuits (see Figure 6-10). The board enables easy access to all the *input / output* (I/O) pins, Port A (RA0 to RA4), and Port B (RB0 to RB7) of the 16F84. The board may also be used with the 16F8X, 16C55X, 16C62X, 16C7X, and 16C8X family of 18-pin PIC microcontrollers.

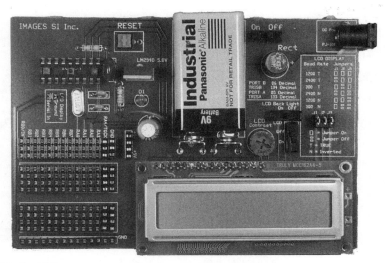

FIGURE 6-10   PIC experimenter's prototype developing board

Its 168-point solderless connection area allows for quick and easy access to all Port A and Port B I/O lines. An open, 18-pin socket can be used for inserting the microcontroller you are developing. The board includes an integrated 16×2 serial LCD that can be easily connected with one wire to any I/O line of the microcontroller (or external source, such as a PC).

Usage

Either an onboard 9-volt battery or an AC/DC transformer can power the experimenter's board. The power switch in the upper right turns power to the board on and off. The board includes a Reset button for resetting the microcontroller. The LCD has its own power switch, located directly above the LCD. It also has a backlight, whose switch is located above the LCD power switch. You can switch on the backlight to conserve power.

The prototyping section of the PIC experimenter's board will be described as the solderless breadboards were, and you will finish up by wiring a simple microcontroller LED project on the experimenter's board. The prototyping area is located at the lower-left corner of the board (see Figure 6-11). An open, 18-pin socket is available to hold the microcontroller being developed.

The prototyping area is similar in design and function to solderless breadboards (see Figure 6-12). You can breadboard (assemble and connect) electronic components and circuits into the prototyping area without soldering. The prototyping area is reusable;

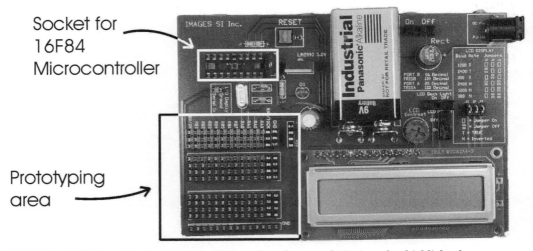

FIGURE 6-11   PIC experimenter's board with breadboard area and 18-pin socket highlighted

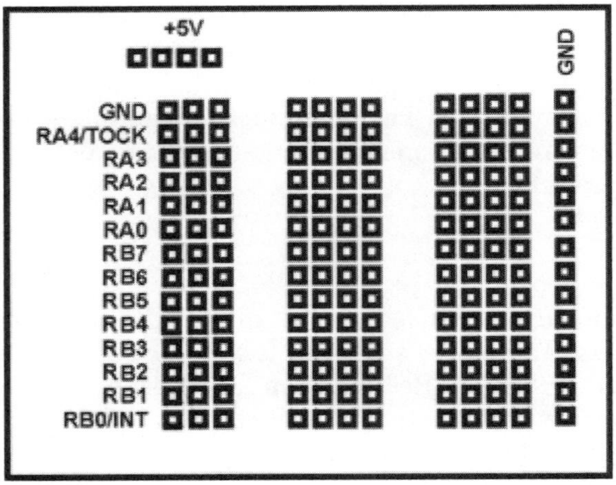

FIGURE 6-12   Diagram of the breadboard area

you can change, modify, or remove circuit components at any time. This makes it easy to correct any wiring errors.

A cutaway of the prototyping area is shown in Figure 6-13. The square holes shown in the figure are sockets. When a wire or pin is inserted into a hole, it makes electrical contact with the underlying metal strip. The holes are spaced so that *integrated circuits*

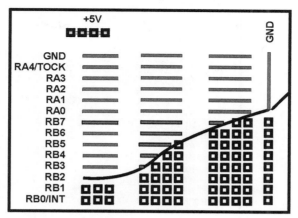

FIGURE 6-13   Cutaway view of breadboard area

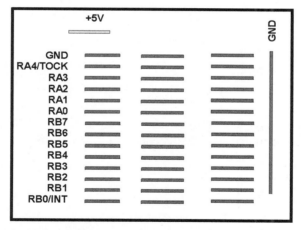

FIGURE 6-14   Underlying electrical connections of the breadboard area

(ICs) and many other components can be plugged right in. The internal electrical connection structure of the prototyping area is shown in Figure 6-14.

In Figure 6-15, you can see at the top of the prototyping area that the columns of Bank 1 are labeled with the pin assignments from the 16F84. These columns are directly connected to those microcontroller pins. Connecting a wire or device to any of the three sockets in a particular column electrically connects that wire or device to the I/O pin of the 16F84.

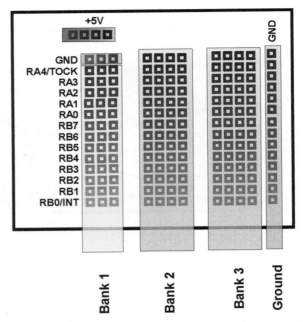

FIGURE 6-15   Breadboard area with banks, ground, and +5-volt power supply highlighted

Bank 2 provides 14 individual, four-socket columns. The four sockets aligned in each individual column are electrically connected. The individual columns are separate electrically from one another.

Bank 3 Is the Same as Bank 2

The last row labeled GND is electrically connected across the entire row. An additional three ground sockets are located at the top of Bank 1, and a +5-volt power is available from a four-socket column adjacent to Bank 1.

Simple Experiment

We shall wire a simple experiment to illustrate the usage of the experimenter's prototyping area for blinking an LED. Yes, this is similar to the wink program, with the exception that you are only using one LED this time. The following code uses a small PICBasic program and a PICBasic Pro program to blink an LED on pin RB1.

```
PICBasic Program          PICBasic Pro Program
start: High 1             start: High PortB.1
Pause 250                 Pause 250
Low 1                     Low PortB.1
Pause 250                 Pause 250
Goto start                Goto start
```

The complete schematic for this experiment is shown in Figure 6-16. Aside from a programmed 16F84, you only need two other components: a 470-ohm, 1/4-watt resistor and a subminiature LED. All the other components needed to make the 16F84 work are already hardwired on the PIC experimenter's board.

The LED has two terminals, one longer than the other. The longer terminal on the LED is positive, shown in the legend of Figure 6-17. On the schematic, the LED appears as a diode. To wire this circuit, connect one lead of the 1/4-watt resistor into one of the RB1 sockets. Connect the other lead of the 1/4-watt resistor into a socket in Bank 2. Take the positive lead of the LED and plug it into a socket in the same column as the one containing the resistor lead. Connect the opposite lead of the LED and plug it into one of the ground sockets at the bottom.

Plug the programmed 16F84 microcontroller into the 18-pin socket on the PIC experimenter's board and turn on the power. The LED should begin blinking on for 1/4 second and then off for a 1/4 second. This on-off cycle (blinking) continually repeats.

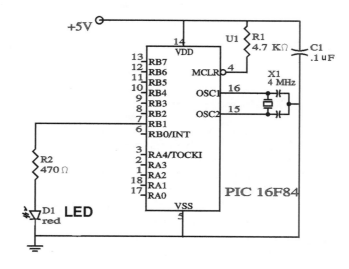

Schematic - LED Blink on RB1

FIGURE 6-16 Schematic of a blink circuit

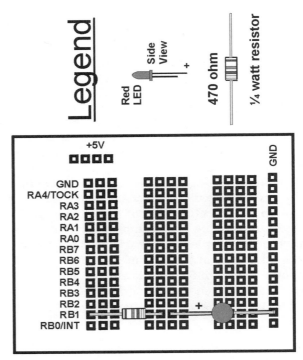

FIGURE 6-17   Blink circuit assembled in the breadboard area of the PIC experimenter's board

Using the X-Board's LCD: PICBasic and PICBasic Pro Examples

The LCD module on the PIC experimenter's board uses the same serial interface and a 2-line by 40-character format (with just 16 characters being displayed per line) as the serial LCD module (see Figure 6-18).

To use the LCD, connect a jumper from the desired output pin on the microcontroller to the LCD Serial In on the experimenter's board. It is not necessary to connect a ground line to the serial input ground unless the serial data is coming from an external source off the experimenter's board.

Data received via the Serial In line appear on the screen. Send the string "Images," and "Images" will appear on the LCD. This LCD obeys the LCD commands described in Chapter 14.

You may need to adjust the contrast control to enhance the display. The contrast control is set fully clockwise (the highest contrast) at the factory, but you can manually adjust the contrast.

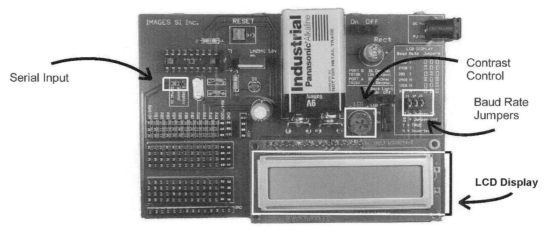

FIGURE 6-18 PIC experimenter's board with LCD controls highlighted

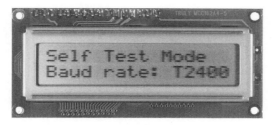

FIGURE 6-19 LCD in self-test mode

The baud rate is set with the onboard jumpers J1 to J3. Set the jumpers in accordance with the silkscreen diagram on the experimenter PC board. At all baud rates, serial data is received at 8 data bits, 1 stop bit, no parity.

The LCD module must match the baud rate and mode of the serial data it will receive from the computer/microcontroller in order for the information to be displayed. The onboard LCD also has a self-testing mode that will print the current baud rate as determined by the jumper settings and mode (true/inverted), as shown in Figure 6-19. To enter self-test mode, connect the serial in line to ground (for true) or +5V (for inverted) upon LCD module startup.

Connect the serial input of the LCD to PortB.1 of a PIC microcontroller. You can use the previous PICBasic or PICBasic Pro LCD programs for testing.

PIC 16F84 Microcontroller

This chapter will look at the specifications and programming of the PIC 16F84 microcontroller. You will learn how to access and configure the microcontroller's *input/output* (I/O) ports by writing into the microcontroller's registers. What you'll learn in this chapter is also applicable to other PIC microcontrollers. Although it appears that the focus is just on the PIC16F84 microcontroller, keep in mind that it is representative of the other PIC microcontrollers as well.

The latest version of the PIC 16F84 from Microchip is the 16F84A. This is essentially the same microcontroller with the same specifications. The die used by Microchip to make the previous version of the microcontroller has been revised, hence the "A" revision. So the 16F84 and 16F84A are used interchangeably throughout the text and are considered to be one and the same.

You may question why I am using the PIC 16F84A as the main microcontroller in this book when more advanced microcontrollers are available on the market for less money. The reason is that is microcontroller is readily available, easy to program, and versatile. In addition, since the venerable 16F84 and 16F84A have been around for quite some time now, a large quantity of programs have been written for them. If you look around on the Internet, you can find many programs written for the 16F84.

Advanced PIC Microcontrollers

After you gain some programming experience with the 16F84, you may want to try a few advanced microcontrollers for the additional features they offer. What are the advantages of this? Take the PIC 16F628 microcontroller, for example. This microcontroller comes in an 18-pin package just like the 16F84A and is a few dollars less. So, not only is this microcontroller less expensive, it has numerous technological advantages, such as

- 2K memory space (as compared to the 16F84A's 1K)
- 16 I/O pins (as compared to the 16F84A's 13 I/O pins)

- 224×8 data *random access memory* (RAM) (twice as much working RAM)
- Analog comparators
- Internal or external oscillator

Still other PIC microcontrollers incorporate more features such as analog-to-digital converters, *universal asynchronous receiver transmitter* (UARTs), more working RAM, greater *electrically erasable programmable read-only memory* (EEPROM), and additional I/O lines.

This book will use the 16F84A microcontroller because it is easy to program with less required setup than the more advanced microcontrollers on the market.

Back to the 16F84 Microcontroller

Before moving into the programming aspects, let's look at some specifications of the PIC16F84 microcontroller (see Table 7-1).

Other factors such as I/O pin loading, the operating voltage, and frequency will have an impact on the operating current. For example, the power-down current in sleep mode (I/O pins at a high-impedance state) is 7 uA.

Clock Oscillators

PIC microcontrollers can be operated in four different oscillator modes. The oscillator mode is selected when programming the microcontroller using the EPIC software. You have the option of selecting one of the following modes:

LP	Low-power crystal
XT	Crystal/resonator
HS	High-speed crystal/resonator
RC	Resistor/capacitor

In the XT, LP, or HS modes, a crystal or ceramic resonator is connected to the OCS1/CLKIN and OSC2/CLKOUT pins to establish oscillation (see Figure 7-1). For crystals 2.0 to 10.0 MHz, the recommended capacitance for C1 and C2 is in the range of 15 to 33 pF. Crystals provide accurate timing to within 50 *parts per million* (PPM). For a 4 MHz crystal, this works out to 200 Hz.

TABLE 7-1   Current maximums for I/O ports

Maximum output current sourced by any I/O pin	20	mA
Maximum input current sunk by any I/O pin	25	mA
Maximum current sourced by Port A	50	mA
Maximum current sunk by Port A	80	mA
Maximum current sourced by Port B	100	mA
Maximum current sunk by Port B	150	mA
Typical operating current	1.8	mA

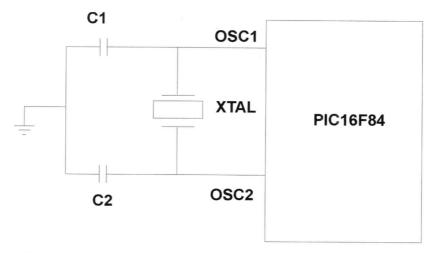

FIGURE 7-1   Microcontroller using a crystal oscillator

A ceramic resonator with built-in capacitors is a three-terminal device that is connected as shown in Figure 7-2. The timing accuracy of resonators is approximately 0.5 percent. For a 4 MHz resonator, this works out to 20,000 Hz.

RC oscillators may be implemented with a resistor and capacitor (see Figure 7-3). Although additional cost saving is provided, applications using *resistor-capacitors* (RC) must be insensitive to timing. In other words, it would be hard to establish RS-232 serial communication using an RC oscillator because of the variance in component tolerances.

To insure maximum stability with RC oscillators, Microchip recommends to keep the R value between 5 and 100 kilo-ohms. The capacitor value should be greater than 20 pF.

No standard formula exists for calculating the RC values needed for a particular frequency. Microchip provides this information in the way of graphs given in the data sheets for particular microcontrollers.

The RC oscillator frequency, divided by four, is available on the OSC2/CLKOUT pin. This output can be used for testing and as a clock signal to synchronize other components.

An external clock may also be used in the XT, LP, or HS modes. External clocks only require to be connected to the OSC1 pin (see Figure 7-4). This is useful when attempting to design an entire circuit that can be implemented with one external clock for all components. Clock accuracy is typically similar to the accuracy quoted for the crystals.

Reset

The PIC16F84 can differentiate between different kinds of resets. During reset, most registers are placed in an unknown condition or a "reset state." The exception to this is a *watchdog timer* (WDT) reset during sleep, because the PICBasic compiler automatically

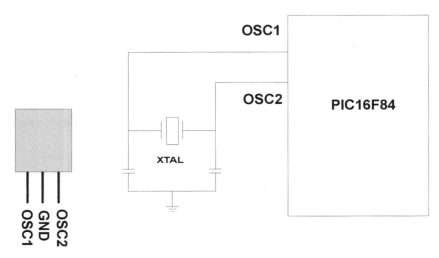

FIGURE 7-2   Microcontroller using a ceramic oscillator

stores a TRISB register when using the Sleep command and reinitializes TRISB after a reset for the resumption of normal operation. The following are reset commands the compiler recognizes:

Power-on reset

MCLR reset during normal operation

MCLR reset during sleep

WDT reset during normal operation

WDT wake-up (during sleep)

For the time being, you will only be concerned with the MCLR reset during normal operation, and the MCLR pin is kept high during such an operation. In the event that it is necessary to reset the microcontroller, bring the MCLR pin momentarily low (see Figure 7-5).

In some cases, you may want to include an optional resistor, R2 (100 ohms). This resistor limits any current flowing into MCLR.

PIC Harvard Architecture

PIC microcontrollers use *Harvard architecture*, which simply means that the memory on the PIC microcontrollers is divided into program memory and data memory. Harvard architecture uses separate buses to communicate with each memory type. It has an improved bandwidth because both memories can be accessed during the same clock instruction, making it faster than the standard *von Neumann architecture,* which uses

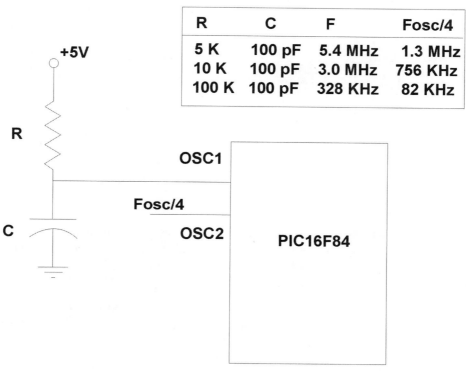

R	C	F	Fosc/4
5 K	100 pF	5.4 MHz	1.3 MHz
10 K	100 pF	3.0 MHz	756 KHz
100 K	100 pF	328 KHz	82 KHz

Vdd = 5 Volts

FIGURE 7-3   Microcontroller using an RC oscillator

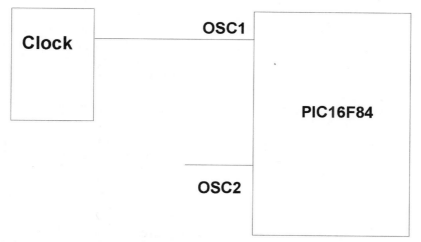

FIGURE 7-4   Microcontroller using a external clock

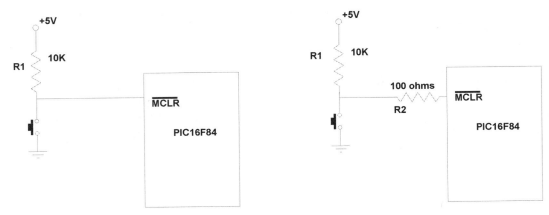

FIGURE 7-5   Connecting a reset switch to microcontroller

a single amount of memory for programs and data. The von Neumann architecture accesses data and memory over the same bus, whereas the Harvard architecture allows for other enhancements. For instance, instructions may be sized differently than 8-bit-wide data. The chip's architecture may be seen in Figure 7-6, which is a block diagram of the 16F84.

The user program memory space extends from 0x0000h to 0x03FFh (0 to 1023 decimal). Accessing a memory space above 03FFh will cause a wraparound to the beginning of the memory space.

Register Map

The register map is a memory area partitioned into two spaces called banks. In Figure 7-7 you can see the two banks: Bank 0 and Bank 1. The small *h* with the numbers under the file address informs us that these are hexadecimal numbers. If you were programming in machine or assembly language, you would have to set a bit in the status register to move between the two banks. Fortunately for you, the PICBasic and PICBasic Pro compilers handle this bank switching task for you.

At this point, two addresses should be mentioned. The first is the reset vector at 00h. Upon power-up or reset, the program counter is set to the memory location held at 00.

The interrupt vector is shown as FSR 04h. Upon an interrupt, the return address is saved and the program execution continues at the address held in this memory location. On a return from interrupt, the program execution continues at the return address previously saved.

Memory Mapped I/O

The data memory in the PIC microcontroller can be broken down further into general-purpose RAM and the *special function registers* (SFRs).

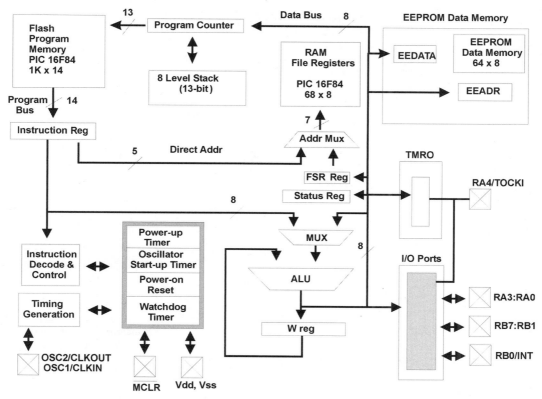

FIGURE 7-6   Block diagram of the 16F84 microcontroller

The registers on the PIC16F84 are mapped in the data memory section at specific addresses. The PICBasic and PICBasic Pro compilers enable you to read and write to these registers as if they are standard memory bytes in RAM. This is an important concept to remember. You can also read and write to the SFR registers as if they are memory locations. The PICBasic compiler uses the commands Peek (read) and Poke (write). The PICBasic Pro compiler can read and write to the registers directly.

By writing numbers into the chip's registers, you program the chip I/O (via the registers) and have it perform the functions you need.

Although you can read and write to the registers using your familiar decimal numbers, to understand what happens inside the microcontrollers registers with those numbers requires a fundamental understanding of the binary number system.

Binary Fundamentals

To access the PIC chip registers efficiently, understanding a little binary goes a long way. Binary isn't difficult to learn because there are only two values. That's what the

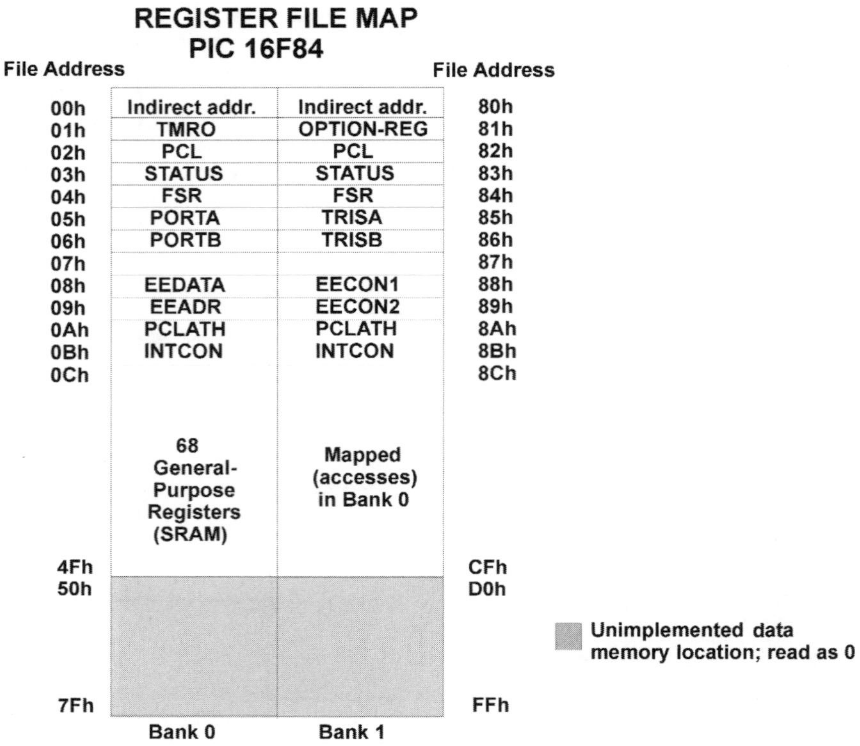

REGISTER FILE MAP
PIC 16F84

File Address			File Address
00h	Indirect addr.	Indirect addr.	80h
01h	TMRO	OPTION-REG	81h
02h	PCL	PCL	82h
03h	STATUS	STATUS	83h
04h	FSR	FSR	84h
05h	PORTA	TRISA	85h
06h	PORTB	TRISB	86h
07h			87h
08h	EEDATA	EECON1	88h
09h	EEADR	EECON2	89h
0Ah	PCLATH	PCLATH	8Ah
0Bh	INTCON	INTCON	8Bh
0Ch			8Ch
	68 General-Purpose Registers (SRAM)	Mapped (accesses) in Bank 0	
4Fh			CFh
50h			D0h
7Fh			FFh
	Bank 0	Bank 1	

Unimplemented data memory location; read as 0

FIGURE 7-7   Register file map for 16F84

word binary means: "based on two," as in the two numbers 0 and 1. The binary 0 and 1 can also be compared to an electrical signal controlled by a switch that has two values: off (0) and on (1). In binary, a digit is called a bit, which stands for *binary digit*. A byte is a digital number containing 8 bits.

An 8-bit byte can hold any decimal value between 0 and 255. In hexadecimal notation, these same values (0 to 255) are expressed as 00 to FF. You are not going to be learning the hexadecimal number system (or hex) primarily because you don't need to use hexadecimal notation to write Basic programs (although you can). It's nice to know hexadecimal notation in the grand scheme of things because it is commonly used when dealing with microcontrollers, and many other writers will use hexadecimal numbers, but it's not essential. What is essential at this point is to gain an understanding of binary, so let's stay focused on this. If you understand the binary number system completely and are still interested in hexadecimal notation, a quick hexadecimal tutorial is included in the appendix.

The *central processing unit* (CPU) in the PIC 16F84A uses an 8-bit data bus (pathway). The registers in the PIC chip are also 8 bits wide. So, a byte is the perfectly sized number to access the PIC chip registers. You will read from and write to the PIC microcontroller registers using the decimal numbers between 0 and 255 that can be contained in one 8-bit byte.

However, when you write a decimal number into a register, the microcontroller only sees the binary equivalent of that decimal number (byte) you wrote to the register. For you to understand what's happening inside the register, you need to be able to look at the binary equivalents of the decimal (byte) number also. Once you can do this, your ability to effectively and elegantly program the PIC microcontrollers is greatly enhanced.

Table 7-2 shows all the decimal and binary number equivalents for numbers 0 through 32. Using this information, the binary equivalent numbers from 32 to 255 can be extrapolated.

Looking at the table, each decimal number on the left side of the equal sign has its binary equivalent on the right side. So when you see a decimal number, the microcontroller will see the same number as a series of eight *bits* (eight bits are in a byte).

Registers and Ports

The PIC 16F84 has two I/O ports: Port A and Port B. Each port has two registers associated with it: the Tri-State Enable (TRIS) register and the port address itself.

The TRIS register is a one-byte (8-bit) programmable register on the PIC 16F84 that controls whether a particular I/O pin on its associated port is configured as an input or output pin. A TRIS register exists for each port. TRISA controls the I/O status for the pins on Port A, and TRISB controls the I/O status for the pins on Port B. Incidentally, the words pins and lines should be taken to mean the same thing and are used interchangeably throughout the text.

Once a port is configured using the TRIS register, the program (user) may then read or write information to the port using the port address.

TABLE 7-2 Binary number equivalents

0 00000000	16 = 00010000	32 = 00100000
1 = 00000001	17 = 00010001	.
2 = 00000010	18 = 00010010	.
3 = 00000011	19 = 00010011	.
4 = 00000100	20 = 00010100	64 = 01000000
5 = 00000101	21 = 00010101	.
6 = 00000110	22 = 00010110	.
7 = 00000111	23 = 00010111	.
8 = 00001000	24 = 00011000	128 = 10000000
9 = 00001001	25 = 00011001	.
10 = 00001010	26 = 00011010	.
11 = 00001011	27 = 00011011	.
12 = 00001100	28 = 00011100	255 = 11111111
13 = 00001101	29 = 00011101	
14 = 00001110	30 = 00011110	
15 = 00001111	31 = 00011111	

On Port B, eight I/O lines are available, whereas Port A only has five I/O lines available to the user. Figure 7-8 shows the relationship between a binary number and the two PIC microcontroller registers that control Port B. Look at the binary numbers shown in Table 7-3. Notice for each progression of the binary 1 to the left, the exponential value of the decimal number 2 is increased by one. This is also shown in Figure 7-8 in the Power of 2 column.

Table 7-3 identifies relevant bit locations (Bit 0, Bit 1 . . . Bit 7), the equivalent decimal number for that bit location, and the binary equivalent for each number. Each progression of the binary 1 to the left identifies a bit location and bit weight (the decimal number equivalent) within the byte (8 bits).

For instance, suppose you wanted to write binary 1's at the RB7 and RB4 locations. To do so, you would add their bit weights together, in this case 128 (RB7) and 16 (RB4),

TABLE 7-3   Binary number progression

Bit #	Decimal		Binary	Bit #	Decimal		Binary
Bit 0	1	=	00000001	Bit 4	16	=	00010000
Bit 1	2	=	00000010	Bit 5	32	=	00100000
Bit 2	4	=	00000100	Bit 6	64	=	01000000
Bit 3	8	=	00001000	Bit 7	128	=	10000000

Port B

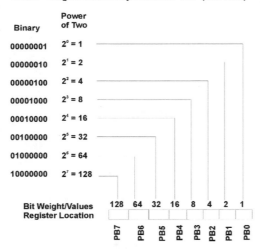
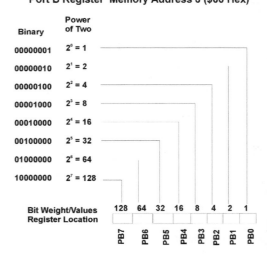

FIGURE 7-8   TRISB register and Port B

which equals 144. The binary equivalent of decimal number 144 is 10010000. If you slide the binary number 10010000 into the open register illustrated in Figure 7-2, you will see that the binary 1's are in the RB7 and RB4 positions. Remember this; it is important.

The open TRISB register shown in Figure 7-8 may be used to examine the function of any number placed in the TRISB register. In the same way, the Port B register may be used to examine any number placed in the Port B register.

The correlation between the bit number, bit weight, and the I/O line is used to program and control the port. A few examples will demonstrate this relationship.

Using the TRIS and Port Registers

If you place a binary 0 at a bit location in TRISB, the corresponding pin location on Port B will become an output pin. If you place a binary 1 at a bit location in TRISB, the corresponding pin on Port B becomes an input pin. The TRISB memory address is 134 (or 86h in hex). After Port B has been configured using the TRISB register, the user can read or write to the port using the Port B address (decimal number 6).

For example, suppose you want to make all Port B lines output lines. To do this, you need to put a binary 0 in every bit position in the TRISB register. The decimal number you would write into the TRISB register is 0. Doing so will make all the Port B I/O lines configured as output lines.

If you connect a *light-emitting diode* (LED) and a current-limiting resistor to each output line, you will see a visual indication of any number you write to Port B. To turn on the LEDs connected to RB2 and RB5, you will need to place a binary 1 at each bit position on the Port B register. To accomplish this, look at the bit weights associated with each line. RB2 has a bit weight of 4, and RB5 has a bit weight if 32. Add these numbers together (4 + 32 = 36) to get 36 and write that number into the Port B register. When we write the number 36 into the Port B register, the LEDs connected to RB2 and RB5 will light.

A similar procedure is used to configure Port A. Begin by using the TRISA register, decimal address 133 (see Table 7-4). On Port A, however, only the first five bits of the TRISA and their corresponding I/O lines on Port A (RA0 through RA4) are available for use (see Figure 7-9). Examine the data sheets from Microchip on the 16F84 to corroborate this information.

TABLE 7-4

Register	Memory location (hexadecimal)	Memory location (decimal)
Port A	05h	5
Port B	06h	6
TRISA	85h	133
TRISB	86h	134

Port A

TRISA   Decimal 133   85 Hex

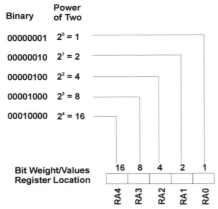

Port A   Decimal 5   05 Hex

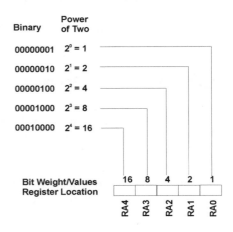

FIGURE 7-9   TRISA register and Port A

On power-up and reset, all the I/O pins of Port A and Port B are initialized (configured) as input pins.

Writing to a Register Using PICBasic Compiler

Using the PICBasic compiler, the command to write to a register is the Poke command. As an example, let's configure Port B so that bit 7 (RB7) is an input pin and all other pins are output lines. To place binary 0's and 1's in the proper bit locations, use the bit weights shown in Figure 7-8. For instance, to turn bit 7 on (1) and all other bits off (0), we would write the decimal number 128 into TRISB for Port B. The program line to write the decimal value 128 into the TRISB register will look like this:

```
Poke 134,128
```

The first number after the word Poke is the memory address the command will write to, in this case 134, which is the data memory address of TRISB for Port B. The next number after the comma is the value we want to write in that memory address. In this case, it's 128.

Writing to a Register Using PICBasic Pro Compiler

Using the PICBasic Pro compiler, the command to write the decimal value of 128 to the TRISB register is

```
TRISB = 128
```

Using the PICBasic Pro, you don't need to know the memory location of the TRISB register; the compiler does this work for you. The names of the microcontroller's SFR registers are already defined for your use.

Look at the binary equivalent of the decimal number 128:

```
1 0 0 0 0 0 0 0
```

If you mentally place each 1 and 0 into the TRISB register locations shown in Figure 7-8, you'll see how the 1 fits into the bit 7 place, making that corresponding line an input line. All other bit locations have a 0 written in them, making them output lines.

To summarize, by writing a decimal number into the TRIS register, the representative binary equivalent of that number that contains a sequence of bits (0's and 1's) configures the pins on the corresponding port to be either an output or input in any combination you require. A binary 1 in any bit locations turns that corresponding bit/pin on the port to an input pin. Likewise, writing a binary 0 into the bit location will turn the corresponding bit/pin on the port into an output pin. In addition, the configuration of the port can be changed "on-the-fly" as the program is running.

The PICBasic compiler uses the Poke command to write to registers. The PICBasic Pro compiler has the registers predefined and may be written to directly.

Accessing the Ports for Output

Once the port(s) lines have been configured using the TRIS register, you can start using it. To output a binary number at the port, write the number to the port. To write to the port using the PICBasic compiler, use the Poke command. To write to a port using the PICBasic Pro compiler, which has the port names and address already predefined, write to the port directly.

The binary equivalent of the decimal number will be output as shown in our first example. To output a high signal on RB3 using the PICBasic compiler, use this command:

```
Poke 6, 8
```

where 6 is the memory address for Port B and 8 is the decimal equivalent of the binary number (00010000) we want to output. Reading input information on the ports will be discussed in Chapter 8, "Reading I/O Lines."

To output a high signal on RB3 using the PICBasic Pro compiler, use either of the two following commands:

```
PortB.4 = 1
```

This command singles out bit 4 on port B (PortB.4) and brings it high (=1).

The next command accesses all the bits on Port B at once:

```
PortB = 8
```

As stated previously, the number 8 is the decimal equivalent of the binary number we want to output.

Electrical Binary, TTL, and CMOS

When a pin on Port-B (RB0 to RB7) is configured as an input line, the microcontroller can read the electrical voltage present on that input pin to determine its binary value (0 or 1).

When a pin on a port is configured as a output, the microcontroller can raise the voltage on that pin to +5 volts by placing a binary 1 at the bit location on the port. A binary 0 at the bit location will output a zero voltage.

When a pin (or bit) is set to 1, it may be called "on," "set," or "high." When a bit is set to 0, it may be called "off," "cleared," or "low."

In *Time to Live* (TTL) logic, electrically a binary 1 is equal to a positive voltage level between 2 and 5 volts. A binary 0 is equal to a voltage of 0 to 0.8 volts. Voltages between .8 and 2 volts are undefined.

The *complementary metal oxide semiconductor* (CMOS) has a slightly different definition. Input voltages within 1.5 volts of ground are considered binary 0, whereas input voltages within 1.5 volts of the +5-volt supply are considered binary 1.

Digital logic chips (TTL and CMOS) are available in a number of subfamilies: CMOS; 4000B, 74C, 74HC, 74HCT, 74AC, 74ACT; and TTL logic chips 74LS, 74ALS, 74AS, and 74F. These differences become important when you need to make different logic families talk to one another.

CMOS devices swing their output rail to rail, so +5-volt CMOS can drive TTL, *negative-channel metal-oxide semiconductor* (NMOS), and other +5-volt-powered CMOSs directly. (The exception to this is old-fashioned CMOS [4000B/74C].) TTL devices, on the other hand, may not output sufficient voltage for a CMOS device to see a binary 1 or a high signal.

This could have been a problem, since the PIC 16F84 is a CMOS device. The designers of the PIC microcontrollers were thoughtful enough to buffer the I/O lines with TTL buffers, thus allowing the PIC I/O lines to accept TTL input levels while outputting full CMOS voltages. This enables you to directly connect TTL logic devices as well as CMOS devices to your PIC microcontroller without difficulty.

Counting Program

To illustrate many of these concepts, I have written a simple program. The schematic for the program is shown in Figure 7.10. It is a binary counting program that will light eight LEDs connected to Port B's eight output lines.

The counting program will light the LEDs in the sequence shown in the binary number table. Each binary 1 in a number in the table will be represented with a lit LED. Every 250 milliseconds ($1/4$ second), the count increments. After reaching the binary number 255 (the maximum value of a byte), the sequence repeats, starting from zero.

Counting in Binary by One

Enter the following program into CodeDesigner exactly as it is written:

```
'PICBasic Program 7.1 -- Binary Counting --
'Initialize variables
Symbol TRISB = 134      'Assign TRISB for Port B to decimal value of 134
Symbol PortB = 6        'Assign Variable PortB to decimal value of 6
'Initialize Port(s)
poke TRISB,0            ' set port B pins to output
Loop:
for B0 = 0 to 255
poke PortB, B0          'Place B0 value at port to light LEDs
pause 250               'Without pause counting proceeds too fast to see
next B0                 'Next B0 value
goto loop
'end
```

The following program is written for the PICBasic Pro compiler:

```
'PICBasic Pro Program 7.2 -- Binary Counting --
'Initialize variables
B0 var byte
'Initialize Port(s)
TRISB = 0               ' set port B pins to output
Loop:
for B0 = 0 to 255
PortB = B0              'Place B0 value at port to light LEDs
pause 250               'Without pause counting proceeds too fast to see
next B0                 'Next B0 value
goto loop
'end
```

Let's look at the program and decipher it line by line. The first two lines are comments that begin with a quote mark (').

```
' Program Binary Counting
' Initialize variables
```

The compiler ignores all text following a quote mark. You should use comments liberally throughout your basic code to explain to yourself what you are doing and how you are doing it. What appears so obvious to you when you are writing a program becomes obscure a few months later. All comments are stripped when the program is compiled into HEX and ASM files, so add as many comments as you like; they do not take up any program space.

In the PICBasic program, the following two lines initialize two important variables. TRISB is assigned the decimal value of 134 and Port B represents the Port B address, or the decimal value of 6, for subsequent use in the program. Technically, you don't need to initialize these variables. You could write the decimal equivalent (number 134) instead of using the TRISB variable when needed by the program. So if you wanted, you could write POKE 134, XX instead of POKE TRISA, XX. However, by initializing variables, especially in more complex programs, using a mnemonics variable for important register addresses makes writing the programs and following the logic easier and less error prone.

```
Symbol TRISB = 134    'Assign TRISB (Port-B) to 134
Symbol PortB = 6      'Assign Variable PortB to decimal value of 6
```

The variable TRISB now represents a decimal value of 134, and the variable Port B now represents a decimal value of 6. Hereafter in the program, we can refer to the TRISB without needing to remember its numerical value and the same is true for Port B. The comments following each instruction provide valuable information on what each command is doing.

In the PICBasic Pro Program, you don't need to define TRISB or PortB; these registers are already defined for you. However, you do need to define the variable B0. Variables in PICBasic Pro may be defined as a bit, byte, or word (two bytes). The variable B0 is defined as a byte, as in the following PICBasic Pro command:

```
B0 var byte    'initialize variable (PICBasic Pro)
'Initialize Port(s)
```

The second line is a comment line that says what follows.

The following line is the command that initializes Port B with a zero, making all of the Port B lines output lines.

PICBasic Compiler:

```
poke TRISB,0    ' set all port B pins to output
```

PICBasic Pro Compiler:

```
TRISB = 0    'set all port B pins to output
Loop:
```

The previous line contains a label called loop. The word loop is clearly identifiable as a label because of the colon mark following the word Loop. Labels can be referred to in the program for jumps (goto's and on value) and subroutines (gosubs).

The following line defines our variable B0. In standard Basic, this line would probably read "for x = 0 to 255." It uses one of PICBasic compiler's predefined variables, B0. The PICBasic Pro compiler does not have predefined variables; this is why the variable needs to be initialized in the PICBasic Pro program.

```
For B0 = 0 to 255
```

Variable Space

The 16F84 has a limited amount of RAM that you can access for temporary storage. In the case of the 16F84, there are 68 bytes of RAM. From this total area of 68 bytes of RAM, 51 bytes are available for user variables and storage.

User Available RAM

RAM may be accessed as bytes (8-bit numbers) or words (16-bit numbers) (see the following table). The PICBasic compiler predefines a number of variables for you. Byte-sized variables are named B0, B1, B2, B3 . . . B51. Word-sized variables are named W0, W1, W2 . . . W25.

The predefined byte and word variables use the same memory space and overlap one another. Word variables are made up of two byte-sized variables. For instance, W0 uses the same memory space of variable bytes B0 and B1. Word variable W1 is made up of bytes B2 and B3 and so on.

The variables B0 and B1 (or W0) in the PICBasic compiler are a little more special than the other variables. The program can read and access the bit status of these two bytes.

Word variables	Byte variables	Bit
W0	B0	Bit 0, Bit 1, . . . Bit 7
	B1	Bit 8, Bit 9, . . . Bit 15
W1	B2	
	B3	
W2	B4	
	B5	
. . .	. . .	
W39	B78	
	B79	

Although the PICBasic Pro does not come with predefined variables, it is easy to define variables using the var command. The advantage of PICBasic Pro variables is that you can read the individual bit status of any variable. With the PICBasic variables, you can only check the bit status of the first two byte variables B0 and B1 (or first word variable W0).

Changing Variable Names (PICBasic Compiler)

Variables may be used for number storage. The variables may also be given a name that has meaning in the program by using the command Symbol. For instance, you could rename your variable B0 to X to make it read more like a standard Basic-language program.

Look back at the PICBasic program. The Symbol command was used at the beginning of the program to store the variables TRISB and Port B. The command can be used by PICBasic Pro users for aliasing variables and constants. It cannot be used to create a variable, but can be used to rename an existing variable.

Overwriting RAM Space

If you write a program that uses more variable space than the PIC microcontroller has RAM to store, the compiler will not generate an error when it compiles the program. However, your program will not function properly, due to the variable space overlapping. So it is up to you to keep track of how many variables are being used in the program. For the 16F84, you may use up to 51 bytes or 25 words, or a combination of both.

When programming other PIC microcontrollers, check their data sheets to see how much RAM they have available.

This next program line writes the value B0 to Port B. Any binary 1's in the number are displayed with a lit LED.

PICBasic Program:

```
Poke PortB, B0     'Place B0 value at port to light LEDs
```

PICBasic Pro Program:

```
PortB = B0
```

The next line simply adds a 1/4-second delay:

```
Pause 250      'Without pause counting proceeds too fast to see
```

This line pauses for 250 milliseconds (1/4 of a second), allowing enough of a time delay to allow you to see the progression:

```
Next B0      'Next B0 value
```

The following line increments the value of B0 and jumps up to the "for B0 = 0 to 255" line. If the value of B0 equals the end value declared in the line (255), the programs drops to the next line.

```
goto loop
```

When B0 equals 255, the for-next loop is finished. The previous line directs the program to jump to the label loop, where the B0 value is reinitialized and the number counting repeats, starting from zero.

Schematic for Program

Figure 7-10 is a schematic with eight LED lights and resistors connected to Port B of a PIC 16F84. Figure 7-11 is a photograph of this project. Notice I used a second solderless breadboard to hold the resistors and LEDs so I wouldn't have to squeeze everything onto a single breadboard.

Counting Binary Progression

The last program was informative. It showed you how to output electrical signals via Port B. Those electrical signals may be used for communication and/or control. As you shall learn in future chapters, an electrical signal off one pin can control just about any household electrical appliance.

However, you are not finished with your simple circuit yet. With a little programming modification, you can have the same circuit perform a binary progression instead of binary counting. What's the difference between a binary progression and counting? The binary progression lights each LED in sequence. It starts with the first LED, then the second, and so on until the last LED is lit; then the progression repeats itself. When the LEDs are arranged in a straight line, the light would appear to travel from one end of the line to the other. If the LEDs were arranged in a circle, the light would travel in a circle.

PICBasic compiler program:

```
'PICBasic Program 7.3 -- Binary Progression Counting --
'Initialize variables
Symbol TRISB = 134 'Assign TRISB B to 134
Symbol PortB = 6 'Assign Variable Port B to decimal value of 6
```

* Capacitors connected to crystals are 22 pF.

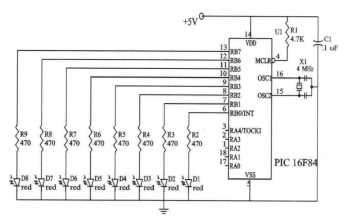

FIGURE 7-10 Schematic for counting program

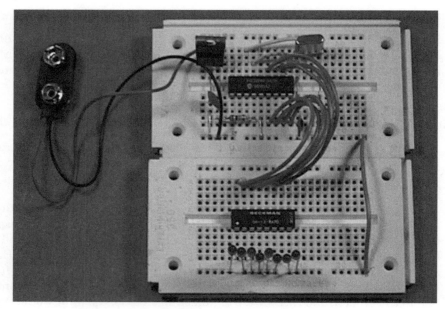

FIGURE 7-11   Counting circuit built on two solderless breadboards

```
'Initialize Port(s)
Poke TRISB,0      ' set port B pins to output
loop:
B0 = 1            ' Set variable to 1 to start counting
B1 = 0            ' Set variable to zero
Poke PortB, B0    'Place B0 value at port to light LEDs
Pause 250         'Without pause proceeds too fast to see
For B2 = 0 to 6
B1 = B0 * 2       'calculate next binary progressive number
B0 = B1           'set B0 to new value
Poke PortB, B0    'Place new value at port to light LEDs
Pause 250         'Without pause counting proceeds too fast to see
Next B2           'Next loop value
goto loop
```

PICBasic Pro compiler program:

```
'PICBasic Pro Program 7.4 -- Binary Progression Counting --
'Initialize variables
B0 var byte
B1 var byte
B2 var byte
'Initialize Port(s)
TRISB = 0         ' set port B pins to output
loop:
B0 = 1            ' Set variable to 1 to start counting
B1 = 0            ' Set variable to zero
PortB = B0        'Place B0 value at port to light LEDs
```

```
Pause 250        'Without pause proceeds too fast to see
For B2 = 0 to 6
B1 = B0 * 2      'calculate next binary progressive number
B0 = B1          'set B0 to new value
PortB = B0       'Place new value at port to light LEDs
Pause 250        'Without pause counting proceeds too fast to see
Next B2          'Next loop value
goto loop
```

Basic High and Low Commands

The way we have defined outputting information thus far is the most powerful and elegant way to do so. However, it's not the easiest. The PICBasic and PICBasic Pro compiler(s) have two commands for outputting information called High and Low.

For PICBasic compiler users, these commands are limited to Port B and will not work on Port A lines. PICBasic Pro compiler users can access other port lines (pins) using these commands.

The High command makes the specified pin output high. The pin so defined is automatically made into an output pin. With the PICBasic compiler, the High command only works with Port B pins 0 to 7. The command structure is as follows:

```
HIGH Pin
```

Here's an example:

```
HIGH 0      'Makes Pin 0 (RB0) an output pin and sets it high (+5V)
```

The PICBasic Pro syntax is backwardly compatible. You may use the same command to control port B pins (lines) 0 through 7. The Pro compiler has the additional feature of being able to access other port lines. For instance, to make Port A, pin 0, an output and set it high, use the following command:

```
High PortA.0
```

You may also use this style to control Port B pins:

```
High PortB.0
```

The Low command makes the specified pin output low. The pin so defined is automatically made into an output pin. With the PICBasic compiler, the Low command only works with Port B pins 0 to 7. The command structure is as follows:

```
LOW PIN
```

Here's an example:

```
LOW 0      'makes Pin 0 an output pin and sets it low(0V)
```

The PICBasic Pro syntax is backwardly compatible, so you can use the same command to control Port B pins (lines) 0 through 7. The Pro compiler has the additional feature of controlling other port pins. For instance, to make Port A, pin 0, an output and set it lower, use the following command:

```
Low PortA.0
```

You can also use this style to control Port B pins:

```
Low PortB.0
```

The High and Low commands are quick and easy commands to use and they do have their usefulness. Real programming power and versatility are obtained using the registers (TRIS and Port) directly. Don't believe it? Try rewriting our simple binary counting programs just using High and Low commands. Call me when you're done.

As a sample program that uses the High and Low commands, here is the first program you worked with:

```
' First Basic program to wink two LEDs connected to PORT B.
Loop:High 0     ' Turn on LED connected to pin RB0
Low 1           ' Turn off LED connected to pin RB1
Pause 500       ' Delay for .5 seconds
Low 0           ' Turn off LED connected to pin RB0
High 1          ' Turn on LED connected to pin RB1
Pause 500       ' Delay for .5 seconds
Goto loop       ' Go back to loop and blink & wink LEDs forever
End
```

Programming Review

Before proceeding to the next chapter, let's take time to review the key programming concepts that have been used in the last few programs.

Comments

Use comments liberally when writing your programs. Use them to describe the logic and what the program is doing at that particular point. This will allow you to follow and understand the program's logic long after you have written (and probably forgotten) the program. Comments begin with a single quote mark (') or with the word REM. The compiler ignores all characters on the line following the quote mark or REM keyword.

Identifiers

Identifiers are names used for line labels and symbols. An identifier may be any sequence of letters, digits, and underscores, but it must not start with a digit.

Although identifiers may be any number of characters in length, the compiler will only recognize the first 32 characters.

Identifiers are not case sensitive, so the labels LOOP:, Loop:, lOOP:, and loop: will be read equivalently.

Line Labels

Labels are anchor points or reference points in your program. When you need the program to jump to a specific program location via Goto, Gosub, or Branch, use a label. Labels are easy to use. Use a descriptive word (identifier) for a label, such as the word "loop:," which we used in programs 7.1 and 7.2. Loop is descriptive in as much as it shows the main loop point for the program. Labels are identifiers followed by a colon (:).

Symbols

Symbols help to make your programs more readable. They use an identifier to represent constants, variables, or other symbols. Symbols cannot be used for line labels.

In your PICBasic programs, you used the symbol TRISB to represent the decimal number 134. The number 134 is the data memory address to the TRISB register for port B. The symbol PortB represents the memory address for Port B. Symbols are easier to remember than numbers. Here are a few examples of the Symbol keyword usage.

```
Symbol  Five = 5          'Symbolic constant
Symbol  Number = W2       'Named word variable
Symbol  Bvalue = BIT0     'Name Bit Variable
Symbol  AKA = Bvalue      'An alias for Bvalue
```

Variables

Variables are temporary storage for your program. For PICBasic compiler users, a number of variables have been predefined for usage in your programs. Byte-sized (8-bit) variables are named B0, B1, B2, and so on. Word-sized (16-bit) variables are named W0, W1, W2, and so on.

PICBasic byte and word variables overlap one another and use the same memory space. The word variables are made up of two byte-sized variables. For instance, the 16-bit W0 variable is made up of the two smaller 8-bit B0 and B1 variables. W1 is made up of B2, B3, and so on. Any of these variables can be renamed to something more appropriate in a program using the Symbol command.

PICBasic variables B0 and B1 (or word variable W0) are special because we can read and test their individual bits (Bit0, Bit1 . . . Bit15). The ability to read the bits in these variables is very attractive for many bit-checking applications. The word variable W0 is composed of the two bytes B0 and B1, and the bit-checking commands will also work with this variable.

PICBasic Pro users need to define all their variables. Variables may be defined as bits (1 bit), bytes (8 bits), or words (16 bits) depending upon your needs. Pro users can read the bit status of all their variables.

Read the specification sheets on the PIC microcontrollers to determine how much free RAM is available. The 16F84 has 68 bytes of free RAM, of which 51 bytes are available to the user.

Next Chapter—Reading Input Signals

The programs you have written thus far have only dealt with outputting binary signals that you can see using the LEDs. Although this is extremely important, it is also just as important to be able to read input off the lines. The status (binary state 0/1) of a line (signal) may be a digital signal or a switch. The next chapter will examine inputting signals to your PIC microcontroller.

Parts List

Same components as outlined in Chapter 1, "Microcontrollers."

Optional Parts

Additional parts include a solderless breadboard, RadioShack PN# 276-175 or an equivalent. It is available from Images Company, James Electronics, JDR MicroDevices, or RadioShack (see suppliers index).

Reading I/O Lines

In the last chapter, you studied outputting binary numbers (information) to Port B and viewing the information using miniature red *light-emitting diodes* (LEDs). In this chapter, you will be inputting and reading binary information from the port(s).

The ability of your microcontroller to read the electrical status of its pin(s) enables the microcontroller to see the outside world. The line (pin) status may represent a switch, sensor, or electrical information from another circuit or computer.

There are three methods (per compiler) you can use to read the electrical status of the microcontroller pins. They are Input, Button, Directly (PICBasic Pro only), and Peek (PICBasic only). Each compiler has one unique method of reading ports. For the PICBasic Pro compiler, you can read a port directly, and with the PICBasic compiler you can use the Peek command.

Placing Electrical Signals on a Pin

In order to learn how to read these electrical signals, you must first test the electrical signals on a pin. Figure 8-1 shows two basic switch schematics, labeled A and B, connected to an *input/output* (I/O) pin. You can use these two switch combinations to put high (binary 1) and low (binary 0) signals on a pin. Let's see how the switches affect the I/O pin electrically.

The switch labeled A in Figure 8-1 connects the I/O pin to a +5-volt power supply through a 10,000-ohm resistor. With the switch open, the electrical status of the I/O pin is kept high (binary 1) at +5 volts. When the switch is closed, the I/O pin connects to ground, and the status of the I/O pin is brought low (binary 0). The 10,000-ohm resistor is there to limit the amount of current (0.5 *milliamperes* [mA]) that could flow into the microcontroller pin from the 5-volt source.

The switch labeled B in Figure 8-1 has an electrical function opposite the switch labeled A. In this case when the switch opens, the I/O pin is connected to ground, keeping the I/O pin low (binary 0). When the switch is closed, the I/O pin is brought high (binary 1) to +5 volts.

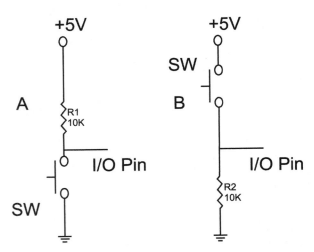

FIGURE 8-1   Switches

These simple switch setups allow you to place high and low signals on a microcontroller pin. As stated previously, these signals can represent electrical signals from any number of sources, including computers, microcontrollers, electrical circuits, or devices. Now we shall begin to look at the diverse commands at your disposal to read the electrical status of the ports and pins on your microcontroller.

Reading a Port

The PICBasic compiler can use the Peek command to check the line status of a port and any pin on that port. As its name implies, the Peek command enables one to view (or peek at) the contents of a specified memory address. In the examples, the memory address "peeked at" is one of the PIC microcontroller's port registers.

An advantage of the Peek command is that it can read multiple lines of a port at once. On the 16F84, Port A has five available I/O lines and Port B has eight available I/O lines. The ability to read multiple input lines at once increases the versatility of your program and allows it to be more concise (less convoluted), shorter, and easier to read.

The Peek command structure is as follows:

```
Peek Address, Var
```

The command is followed by a memory address (to be peeked at), a comma, and a storage variable to hold the peeked value. The peeked valued is stored in the variable VAR after the program line is executed.

In the PICBasic Pro version, you can read the port directly or assign a variable to hold the peeked value. In the following command, a variable is assigned to the value of the port register:

```
Var = PortA
```

After the program line is executed, the variable *Var* contains the peeked value in Port A.

Typically, the variable used in these commands (PICBasic and PICBasic Pro) is a byte-size variable. For instance, using a variable label B0 in PICBasic Pro, we must first initialize the variable using the following command:

```
B0 var byte
```

Then we can use the variable to peek (read) the port using the following command:

```
B0 = PortB
```

In some cases, with PICBasic Pro, you will not need to use an intermediary variable to hold the read value of a port. You can read the port or a individual pin off the port directly in the decision-making program line. For example, in the following line we are checking the status of RA0 off of Port A:

```
If PortA.7 = 1 Then XXX
```

To emphasize the point, let's write a PICBasic and PICBasic Pro program that reads the port. This program uses the A-style switch shown in Figure 8-1. The schematic for the following two programs is shown in Figure 8-2.

This program counts in binary 0 through 15 on pins RB0 to RB3. Pin RB7 is connected to a switch. When the switch is pressed, it brings RB7 down to ground and stops the counting. When the switch is opened again, the counting resumes.

```
'PICBasic Program 8.1
Symbol TRISB = 134        'Set Data Direction Register Port B
Symbol PORTB = 6          'Initialize Port B to 6
'Initialize Port(s)
Poke TRISB,128            'Port B pins (0-6 output) (7 input)
start:                    'Start Program
For B1 = 0 to 15          'Define counting loop
Poke PORTB, B1            'Output current number
Peek PORTB,B0             'Read port
IF bit7 = 0 Then hold     'Is sw1 closed?
check:                    'Label to return from hold routine
Pause 250                 'Slow down to observe counting
Next B1                   'Increment to next number
GoTo start                'counting finished -- start over
hold:                     'If sw1 is press hold
```

* Capacitors connected to crystals are 22 pF.

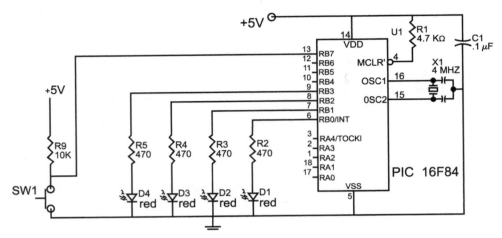

FIGURE 8-2   Schematic of Program 8.1

```
Peek PORTB,B0              'Read sw1 again
IF bit7 = 0 Then hold      'If pressed stay in hold routine
GoTo check                 'Not pressed jump back to counting
```

PICBasic Compiler and Variables B0 and B1 (Bit0 to Bit15)

The first two bytes of RAM memory used for variables B0 and B1 are special. This is because you can test the bit values in these two bytes. If you remember, for byte B0, the bit variables are predefined as Bit 0 through Bit 7. For byte B1, the predefined bit variables are Bit 8 to Bit 15. We test Bit 7 in the program to determine if sw1 is closed.

The following is a PICBasic Pro program equivalent of the previous program. Notice when the status of Bit 7 on Port B is read, you do so directly without the use of an intermediary variable.

```
'PICBasic Pro Program 8.2
B1 VAR BYTE               'Define Byte
'Initialize Port(s)
TRISB = 128               'port B pins (0-6 output) (7 input)
start:                    'Start of program
For B1 = 0 to 15          'Define counting loop
PORTB = B1                'Output current number
IF PORTB.7 = 0 Then hold  'Read port directly - Is sw1 closed?
check:                    'Label (anchor) to return from hold routine
Pause 250                 'Slow down to observe counting
Next B1                   'Increment to next number
GoTo start                'Finished counting - start over
```

```
hold:                            'If sw1 is press hold
IF PORTB.7 = 0 Then hold         'Read port directly, pressed stay in hold routine
GoTo check                       'Not pressed jump back to counting
```

If you wanted, you could have used an intermediary variable in the PICBasic Pro program, but it is not necessary.

Dynamic Changes

In the following programs, you will read the status of two input pins and make dynamic changes in the program as it is running in response to the status of those pins.

The schematic is shown in Figure 8-3. RA0 and RA1 lines off Port A are the two input pins. They are each kept normally high (+5 volts) and are binary 1 through the 10K resistor. When a switch is closed, it connects the pin to ground (binary 0). This project is shown in Figure 8-4.

Using Port A to read your switches frees up the eight lines to Port B, so you can use the entire Port B lines to light eight LEDs. This will allow the program to count up to decimal 255.

One switch connecting RA1 will increase the speed of the binary counting, and the other switch connected to RA0 will decrease the speed of the binary counting.

```
'Program 8.3 For the PICBasic Compiler (PBC)
Symbol TRISB = 134               'Set Data Direction Register Port B
Symbol TRISA = 133               'Set DDR PortA
Symbol PORTB = 6                 'Initialize Port B to 6
Symbol PORTA = 5                 'Initialize Port A to 5
Symbol delay = W2                'Set up delay variable
delay = 250                      'Initialize delay value
```

* Capacitors connected to crystals are 22 pF.

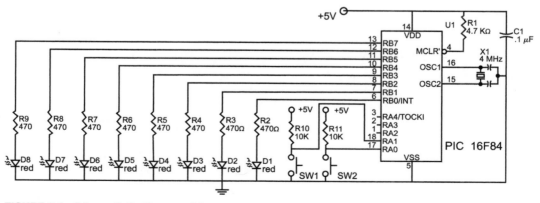

FIGURE 8-3 Schematic for Program 8.3

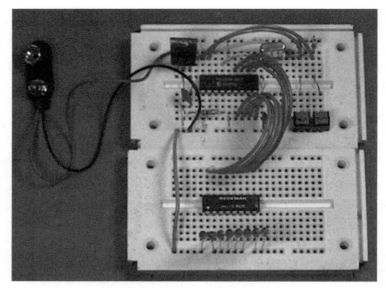

FIGURE 8-4   The configuration for the Dynamic Changes program

```
'Initialize Port(s)
Poke TRISB,0              'set port B pins as output
Poke TRISA,3              'set pin1 & pin2 of port A as input
Loop1:                    'Counting loop
For B1 = 0 to 255         '
Poke PORTB, B1            'Place B1 value at port to light LEDs
Pause delay              'Pause or proceeds too fast to see
Peek PORTA,B0            'Peek SW status on PortA
IF bit0 = 0 Then loop2   'If SW1 closed jump to loop2
IF bit1 = 0 Then loop3   'If SW2 closed jump to loop3
Next B1                   'Next B1 value
GoTo loop1               'repeat
Loop2:                    'Increment binary counting delay
delay = delay + 10        'increase delay by 10 ms
Pause 100                 'delay or timing changes too quickly
IF delay > 1000 Then hold1  'not over 1-second delay
Peek PORTA,B0            'Peek SW1 Status on Port A
IF bit0 = 1 Then loop1   'If opened jump back to loop1
GoTo loop2               'repeat
Loop3:                    'decrement binary counting delay
Peek PORTA,B0            'SW2 Status on Port A
IF bit1 = 1 Then loop1   'If opened jump back to loop1
delay = delay - 10        'decrease delay by 10 ms
Pause 100                 'delay or timing changes too quickly
IF delay < 20 Then hold2  'if less than 20 hold
GoTo loop3               'repeat
hold1:                    'Hold at one second routine
delay = 1000
GoTo loop2
```

```
hold2:                          'Hold at 20 ms routine
delay = 20
GoTo loop3
```

In the following PICBasic Pro Program 8.4, you will see how the enhanced syntax of the Pro compiler helps keep the program concise, easy to read, and easy to understand as compared to Program 8.3. In particular, examine the If . . . Then loops that allow greater program control than the standard If . . . Then loops in the PICBasic compiler. The If . . . Then loops for each compiler are explained in greater detail in Chapter 9, "PICBasic Language Reference," and in Chapter 10, "PICBasic Pro Compiler Additional Commands Language Reference."

```
'Program 8.4 For the PICBasic Pro compiler (PBP)
'Initialize variables needed in program
B2 VAR BYTE
delay VAR WORD                  'Delay variable
delay = 250                     'Initialize delay value
'Initialize Port(s)
TRISB = 0                       'set port B pins as output
TRISA = 3                       'set pin 1 & pin 2 of port A as input
Loop1:                          'Counting loop
For B2 = 0 to 255               '
PORTB = B2                      'Place B2 value at port to light LEDs
Pause delay                     'Pause or proceeds too fast to see
IF PORTA.0 = 0 Then             'Read port directly - If SW1 closed
delay = delay + 10              'increase delay by 10 ms
EndIF
IF delay > 1000 Then            'Is delay over 1 second long
      delay = 1000              'limit to 1 second
EndIF
IF PORTA.1 = 0 Then             'Read Port Directly - If SW2 closed
      delay = delay - 10        'decrease delay by 10
EndIF
IF delay < 20 Then              'If delay less than 20 ms
      delay = 20                'Hold delay at 20 ms
EndIF
Next B2                         'Next B2 value
GoTo loop1                      'repeat
```

Delay Variable

Notice that for the delay variable a two-byte word variable was used. Can you figure out the reason why you need a two-byte variable? If you think it's because a one-byte variable can only hold a maximum number of 255, and your delay can go up to 1,000, you are right. In order to hold a number greater than 255, you need to use at least two bytes. So what is the maximum number your two-byte variable can hold? Answer: 65,535. If you used the maximum delay, your word variable allowed would have to wait over a minute (65.5 seconds) for each increment in the count.

Incidentally, you could have reduced the size of this program by eliminating the lines for the TRISA. If you remember, upon startup or reset all port lines are configured as input lines. Since this is how we need Port A set up, I could have eliminated those lines dealing with the TRISA. Instead, I decided to show a standard Port A setup even though it wasn't needed in this particular application.

Basic Input and Output Commands

In the previous programs, you directly wrote to the PIC microcontroller TRIS registers (A or B) to set various pins on the port to be either input or output lines. However, an easier way exists for configuring pins; the PICBasic and PICBasic Pro compilers have two commands for making pins either input or output lines. The commands are Input and Output. First, I will explain the commands as they relate to the PICBasic compiler, and they only work on Port B pins.

The input line makes the specified pin an input line. Only the pin number itself, that is 0 to 7, is specified (not Pin 0):

```
Input Pin
```

Here's an example:

```
Input 2      'Makes Pin2 an input line.
```

The opposite of the input command is the output command:

```
Output Pin
```

It makes the specified pin an output line. Only the pin number itself, 0 to 7, is specified (not Pin 0).

Here's an example:

```
Output 0      'Makes Pin0 an output line
```

Basic Input and Output Commands (Pro Version)

With the PICBasic Pro, you can access any port pin using the Input and Output commands:

```
Input PortA.0     'Makes Port A, pin 0 an input
Output PortA.1    'Makes Port A, pin 1 and output
```

In addition to accessing additional port lines, the Pro version of these commands is backwardly compatible with the PICBasic compiler. You can use the Input Pin or Output Pin commands in the same manner as described for the PICBasic compiler, where Pin is any number between 0 and 7. When used in this manner, the Input and Output commands will access the Port B line as described previously. The last command we will look at to read input lines is the Button command.

The Button Command

Both the PICBasic and PICBasic Pro compilers have a simple command to read the electrical status of a pin called the Button command. This command, while useful, has a few limitations. With the PICBasic compiler, the Button command may only be used with the eight pins that make up Port B. With PICBasic Pro, it may be applied to read any line off any port. Another limitation is that you can only read a single pin status at a time.

As the name implies, the Button command was implemented to read the status of an electrical "button" switch connected to a port pin (refer to the switches back in Figure 8-1). For the sake of these examples, we will use a line off port B, so the same command can be applied using PICBasic or PICBasic Pro.

The Button command structure is shown here:

```
Button Pin, Down, Delay, Rate, Var, Action, Label
```

The different parts to the command's structure are as follows:

- **Pin**   In PICBasic, this is the pin number (0 through 7) and is for Port B only. In PICBasic Pro, this is any port/pin name (that is, PortA.2).
- **Down**   The state of the pin when the button is pressed (0/1).
- **Delay**   The cycle count before autorepeat starts (0–255). If 0, no debounce or autorepeat is performed. If 255, debounce, but no auto-repeat, is performed.
- **Rate**   The auto-repeat rate (0–255) cycles between auto-repeats.
- **Var**   The byte variable used for the delay/repeat countdown. It should be initialized to 0 prior to use.
- **Action**   The state of the button to perform Goto (0 if not pressed, 1 if pressed).
- **Label**   Execution resumes at this label if Action is true.

Typically, the Button command is used inside a program loop, where the program is looking for a change of state (switch closure). When the state of the I/O pin (line) matches the state defined in the Down parameter, the program execution jumps out of the loop to the "Label" portion of the command line.

Debouncing a Switch

Debounce is a term used to describe the elimination of noise from electrical switches. If you took a high-speed electrical photograph of an oscilloscope's electrical switch closing or opening, the switch's electrical contacts make and break electrical connections many times over a brief period of time (5 to 20 milliseconds). This making and breaking of electrical contacts is called a bounce, because the contacts can be easily visualized as bouncing together and separating. Computers, microcontrollers, and many electronic circuits are fast enough to see this bouncing as multiple switch closures (or openings) and respond accordingly. These responds are typically called *bounce errors*. To circumvent these types of bounce errors, debounce circuits and techniques have been developed. The button command has debounce features built in.

Auto-Repeat

If you press a key on your keyboard, the character is immediately displayed onscreen. If you continue to hold the key down, a short delay occurs, followed by a stream of characters appearing onscreen. The Button command's auto-repeat function can be set up the same way.

Button Example

If you wanted to read the status of a switch off I/O pin 7 on Port B, you would use the following command in the next program. In PICBasic, it would be written as follows:

```
Button 7, 0,254,0,B1,1,loop
```

In PICBasic Pro, the command would be as follows:

```
Button PortB.7,0,254,0,B1,1,loop
```

In actuality, you could use the PICBasic compiler version of the Button command line in a PICBasic Pro program. The Pro version of this command is backward compatible, so that pins 0 through 7 can access the Port B lines just like the PICBasic version of the Button command.

The next program is similar to the counting programs you have already encountered. In this program, you will use PB7 (pin 7) as an input and can use the same schematic as shown in Figure 8-2.

The program contains two loops. The first loop counts to 15, and the current number's binary equivalent is reflected by the lit LEDs connected to port B. The loop continues to count as long as the switch SW1 remains open.

When SW1 is closed, the Button command jumps out of loop 1 into loop 2. Loop 2 is a noncounting loop where the program remains until the switch SW1 is reopened. You can switch back and forth between counting and noncounting states.

```
'PICBasic Program 8.5
Symbol TRISB = 134                 'Set TRIS B to 134
Symbol PortB = 6                   'Set Port B to 6
'Initialize Port(s)
poke TRISB,128                     'set port B pins (1..6) output, Pin 7 input
Loop1:                             'counting loop
for B0 = 0 to 15
poke PortB, B0                     'Place B0 value at port to light LEDs
B1 = 0                             'set button variable to 0:
```

```
pause 250                              'pause counting or it's too fast to see
Button 7,0,254,0,B1,1,loop2            'Check Button status - if closed jump
next B0                                'Next B0 value
goto loop1
loop2:                                 'Second loop Not Counting
B1= 0                                  'Set button variable to zero before use
Button 7,1,254,0,B1,1,loop1            'Check button status - if open jump back
goto loop2
end
* * * * * * * * * * * * * * * * * * * * * * * * * * * * * * * * * * * * * * * * * * * * * * * * * * * * * * * * * * * * * * * * * * * * * * * * * * * * * *
'PICBasic Pro Program 8.6
B0 var byte
B1 var byte
'Initialize Port(s)
TRISB = 128                            'set port B pins (1..6) output, Pin 7 input
Loop1:                                 'counting loop
for B0 = 0 to 15
PortB = B0                             'Place B0 value at port to light LEDs
B1 = 0                                 'set button variable to 0:
pause 250                              'pause counting or it's too fast to see
Button Portb.7,0,254,0,B1,1,loop2      'Check Button status - if closed jump
next B0                                'Next B0 value
goto loop1
loop2:                                 'Second loop Not Counting
B1= 0                                  'Set button variable to zero before use
Button Portb.7,1,254,0,B1,1,loop1      'Check button status - if open jump back
goto loop2
* * * * * * * * * * * * * * * * * * * * * * * * * * * * * * * * * * * * * * * * * * * * * * * * * * * * * * * * * * * * * * * * * * * * * * * * * * * * * *
```

When the program is run, it begins counting, and when the switch is closed, the program stops counting. Open the switch and the counting resumes, starting from 0. You can use the Button commands to produce dynamic changes in the program as you had before.

The Variable Used in the Button Command

The Button command line states that the byte variable used for the delay/repeat countdown should be set initialized to zero prior to use.

Multiple Statements—Single Line

As with the standard Basic language, we can place multiple statements on a single line. The statements must be separated by a colon (:). An example would be B1 = 0: B2 = 0. Here we set the values of variables B1 and B2 to zero.

PICBasic Language Reference

Before you proceed further into PIC microcontroller applications and projects, this chapter is devoted to an overview of the PICBasic compiler language commands. This book deals with two separate but similar compilers: the PICBasic and PICBasic Pro.

Most of the PICBasic compiler commands are common to both compilers, and this chapter lists the PICBasic commands (see Table 9-1). In some cases, the PICBasic Pro compiler version(s) of a command has additional features; these are marked with a single asterisk (*). Commands marked with a double asterisk (**) should not be used with the PICBasic Pro Compiler.

In addition to greater command features, the PICBasic Pro compiler has many more commands at its disposal. The additional PICBasic Pro compiler commands will be provided in the next chapter (Chapter 10, "Additional Command Reference for PICBasic Pro").

Before listing the additional PICBasic Pro commands, we will take a closer look at the syntax of each PICBasic command.

Branch

Branch uses Offset (byte variable) to index into the list of labels. Execution continues at the indexed label according to the Offset value. For example, if Offset is 0, the program execution continues at the first label specified (*Label0*) in the list. If the Offset value is 1, then the program continues at the second label in the list.

```
Branch Offset, (Label0, Label1 . . . LabelX)
```

Here's an example:

```
Branch B8, (label1, label2, label3)
```

TABLE 9-1   PICBasic Commands

Branch *	Computed Goto (equivalent to On . . . Goto).
Button	Input on specified pin.
Call	Call assembly language subroutine at specified label.
Eeprom	Define initial contents of on-chip *electrically erasable programmable read-only memory* (EEPROM).
End	Stop program execution and enter low-power mode.
For . . . Next *	Execute a defined For-Next loop.
Gosub	Call Basic subroutine at specified label.
Goto	Jump program execution to specified label.
High*	Makes specified pin output and brings it high.
I2cin	Reads bytes from I²C device.
I2cout	Writes bytes to I²C device.
If . . . Then *	Compares and uses Goto if specific condition is true.
Input*	Makes specified pin an input.
Let	Performs math and assigns result to variable.
Lookdown *	Searches table for value.
Lookup	Fetches value from table.
Low *	Makes specified pin output and brings it low.
Nap	Powers down the processor for a short period of time.
Output *	Makes the specified pin an output.
Pause*	Delays (1-millisecond resolution).
Peek**	Reads byte from PIC microcontroller register.
Poke**	Writes byte to PIC microcontroller register.
Pot	Reads potentiometer on specified pin.
Pulsin	Measures pulse width (10-usec resolution).
Pulsout	Generates pulse (10-usec resolution).
Pwm	Outputs a pulse-width-modulated signal from the pin.
Random	Generates a pseudo-random number.
Read	Reads a byte from on-chip EEPROM.
Return	Returns from subroutine.
Reverse	Reverses I/O status of pin; input becomes output and vice versa.
Serin*	Asynchronous serial input (8N1).
Serout*	Asynchronous serial output (8N1).
Sleep	Powers down processor (1-second resolution).
Sound	Generates tone or white-noise on specific pin.
Toggle	Makes specified pin an output and toggle state.
Write	Writes byte to on-chip EEPROM.

If B8 = 0, then program execution jumps to label 1.

If B8 = 1, then program execution jumps to label 2.

If B8 = 2 ,then program execution jumps to label 3.

Button

The Button command line consists of the following elements, as shown in the following code line:

Pin	Pin number 0 to 7, Port B pins only.
Down	State of pin when switch is pressed (0 or 1).
Delay	Delay before auto-repeat begins, 0 to 255.
Rate	Auto-repeat rate, 0 to 255.
Var	Byte-sized variable needed for delay repeat. Should be initialized to 0 before used.
Action	State of pin to perform Goto (0 if not pressed, 1 if pressed).
Label	Program execution continues at this label if Action is true.

```
Button Pin, Down, Delay, Rate, Var, Action, Label
```

In the following line, Button checks for the button pressed on pin 0 (Port B) and goes to the label "loop" if it's not pressed.

```
Button 0,0,255,0,B0,0,Loop
```

For PICBasic Pro users, the pin variable may be used with other Port pins besides PORTB, such as PORTA.0. Thus, you can change your previous command to

```
Button PORTA.0,0,255,0,B0,0,Loop
```

which checks for a button pressed on pin PortA.0 and goes to label loop if it's not pressed (see Figure 9-1).

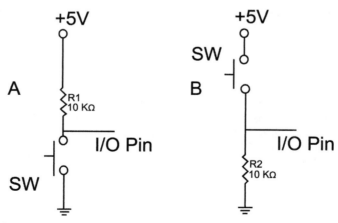

FIGURE 9-1 Switches

Call

The following line jumps to an assembly language routine named Label:

```
Call Label
```

The following line jumps to an assemble language subroutine named storage. Before program execution jumps to the storage routine, the next instruction address after the Call instruction is saved. When the Return instruction is given by the storage routine, the previously saved instruction address is pulled, and program execution resumes at the next instruction after Call.

```
Call storage
```

Eeprom

The eeprom command stores constants in consecutive bytes in on-chip EEPROM. This only works with PIC microcontrollers that have EEPROM such as the 16F84 and 16C84. The command structure is as follows:

```
eeprom     Location, (constant, constant . . . constant)
```

As an example, this command stores 10, 7, and 3 starting at EEPROM location 4:

```
Eeprom 4, (10, 7, 3)
```

End

This command terminates program execution and enters low-power mode by executing continuous Nap commands:

```
End
```

For . . . Next

In the structure of a For . . . Next command, Index is the variable holding the initial value of start, Start is the initial variable, and Step is the value of the increment. If no Step value is specified, it is incremented by 1 each time a corresponding NEXT statement is encountered.

```
For     Index = Start to Stop (Step (-) Inc)
        Body
Next    Index
```

The Step increment value may be positive or negative. If Step and Inc are eliminated, the step defaults to positive 1.

Stop is the final value. When the Index equals Stop, the corresponding Next statement stops looping back to For and execution continues with the next PICBasic statement.

The Body section consists of the basic statements that are executed each time through the loop. The body is optional and may be eliminated as is the case in time delay loops. Here's an example:

```
For B0 = 0 to 127
Poke PortB, B0        'Place B0 value at port to light LED's
Next B0               'Next B0 value
```

Gosub

The Gosub command is structured as follows:

```
Gosub Label
```

Program execution jumps to statements beginning at Label. A Return statement must be used at the end of the label subroutine to return the program execution back to the statement following the Gosub statement.

Gosub statements may be nested, but nesting should be restricted to no more than four levels. An example follows:

```
Gosub wink        'Execute subroutine named wink
 .                'Program execution returns back to here
 .                'Other programming goes here
 .
wink:             'Label wink
High 0            'Bring pin0 high lights LED
pause 500         'Wait 1/2  second
Low 0             'Bring Pin0 low, turns off LED
Return            'Return to main routine
```

Gosub Nesting

Nesting is a term used to describe a second Gosub routine called from a previous Gosub routine. Because of memory limitations, Gosubs can only be nested to a maximum of four levels deep.

Goto

When using the Goto command, program execution jumps to statements beginning at Label:

```
Goto Label
```

Here's an example:

```
Goto loop        'Program execution jumps to statements
                 'beginning at loop.
loop:
for b0 = 1 to 10
poke portB, b0
next
```

High

The High command makes the specified pin an output pin and brings it high (+5 volts). Only the pin number itself, 0 to 7, is specified in the command and it works only on Port B pins:

```
High Pin
```

Here's an example:

```
High 2       'Makes Pin2 (RB2) an output pin and brings it high
             '(+5v)
```

I2CIN

This command allows one to read information from serial EEPROMs using a standard two-wire I²C interface:

```
I2CIN Control, Address, Var (, Var)
```

The second (,Var) shown in the command is only used for 16-bit information. Stored information in a serial EEPROM is nonvolatile, meaning that when the power is turned off the information is maintained. Table 9-2 is a list of compatible serial EEPROMs.

TABLE 9-2   Compatible serial EEPROMs

Device	Capacity	Control	Address Size
24LC01B	128 bytes	01010xxx	8 bits
24LC02B	256 bytes	01010xxx	8 bits
24LC04B	512 bytes	01010xxb	8 bits
24LC08B	1K bytes	01010xbb	8 bits
24LC16B	2K bytes	01010bbb	8 bits
24LC32B	4K bytes	11010ddd	16 bits
24LC65	8K bytes	11010ddd	16 bits

bbb = block select bits (each block = 256 bytes)
ddd = device's select bits
xxx = don't care

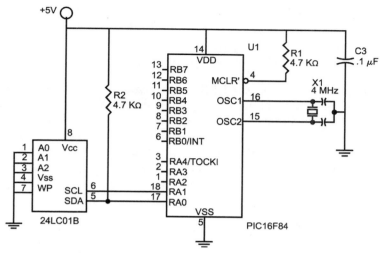

FIGURE 9-2   Schematic of a 24LC01B connected to a PIC 16F84

The high order bit of the Control byte is a flag that indicates whether the following address being sent is 8 or 16 bits. If the flag is low (0), then the address is 8 bits long. Notice EEPROMs 24LC01B to 24LC16bB have the flag set to zero.

The lower seven bits of the Control contain a four-bit control code followed by the chip select or address information. The four-bit control code for a serial EEPROM is 1010. Notice that in all the listed serial EEPROMS this same four-bit control code follows the high bit flag.

The I²C data and clock lines are predefined in the main PICBasic library. The I²C lines are Pin 0 (data) and Pin 1 (clock) of Port A. The I²C lines can be reassigned to other pins by changing the equations at the beginning of the I²C routines in the PBL.INC file. Figure 9-2 is a schematic of a 24LC01B connected to a PIC 16F84.

```
Symbol control = %01010000
Symbol address = B6                'Sets variable address to B6
       address = 32                'Sets address to equal 32
       I2Cin control, address, B2  'Read data from EEPROM
                                   'address 32 into B2
```

NOTE: The PICBasic Pro command equivalent is IC2read.

I2cout

The I2cout command enables one to write information to serial EEPROMs using a standard two-wire I²C interface. The second *(,Value)* shown in the command is only used for

16-bit information. Stored information in a serial EEPROM is nonvolatile, meaning that when the power is turned off the information is maintained.

```
I2cout Control, Address, Value (, Value )
```

When writing to a serial EEPROM, one must wait 10 milliseconds (device dependent) for the Write command to complete before communicating with the device becomes possible. If one attempts an I2cin or I2cout before the Write (10 milliseconds) is complete, the access will be ignored. Using a Pause 10 statement between multiple writes to the serial EEPROM will solve this problem.

The Control and Address are used in the same way as described for the I2cin commands.

```
Symbol  control = %01010000
Symbol  address = B6              'Sets variable address to B6
        address = 32             'Sets address to equal 32
        I2Cout control, address, 16   'Writes data number 16 to
                                      'EEPROM at address 32
        Pause 10                 'Wait 10 ms for write cycle
                                 'to complete.
        address = 33
        I2Cout control, address, 21
        Pause 10
        .
```

NOTE: The PICBasic Pro command equivalent is IC2write.

If . . . Then

This command performs a comparison test. If the particular condition is met (true), then the program execution jumps to the statements beginning at Label. If the condition is not true, program execution continues at the next line.

```
If Comp Then Label
```

The Then in the If . . . Then is essentially a Goto. Another statement cannot be placed after the Then; it must be a label.

The command compares variables to constants or to other variables. If only one variable is used in a comparison, it must be placed on the left. All comparisons are unsigned.

The following is a list of valid comparisons:

=   Equal

>   Greater than

$<$ Less than

$<>$ Not equal

$<=$ Less than or equal to

$>=$ Greater than or equal to

Here's an example:

```
If B8 <= 25 then loop
```

If the value in variable B8 is less than or equal to 25, then the program jumps to loop. The If . . . Then command may also be used with two binary logic comparisons, And and Or.

The PICBasic Pro version of this command has additional features.

Input

This command makes the specified pin an input pin. Only the pin number itself, 0 to 7, is specified in the command. Input works only on Port B pins. Its structure is as follows:

```
Input Pin
```

Here's an example:

```
Input 1      'Makes Pin1 (RB1) an input.
```

Let

The Let command assigns a value to a variable:

```
Let Var = Value
```

You can also optionally use the following command:

```
Where Value = OP Value
```

The value assigned may be

- A constant (Let B1 = 27)
- The value of another variable (Let B1 = B2)
- The result of one or more binary (math) operations

TABLE 9-3   Valid operations

+	Addition
−	Subtraction
*	Multiplication
**	MSB of multiplication
/	Division
//	Remainder
MIN	Minimum
MAX	Maximum
&	Bitwise AND
|	Bitwise OR
^	Bitwise XOR
& /	Bitwise AND NOT
| /	Bitwise OR NOT
^ /	Bitwise XOR NOT

The operations are performed strictly left to right, and all operations are performed with 16-bit precision (see Table 9-3).

Here are some sample operations:

```
Let B1 = 34          'Assign variable B1 the value of 34
                     '("Let" is optional)
Let B1 = B0 / 2      'Assign variable B1 to B0's value shifted right
                     'one bit (divided by 2)
```

When multiplying two 16-bit numbers, the results are the lower 16 bits of the 32-bit result:

```
Let W1 = W0 * 256    'Multiply value held in W0 by 256 and
                     'place resultant in W1 (lower 16 bits)
```

If one requires the higher-order 16 bits, use the following command:

```
Let W1 = W0 ** 256   'Multiply value held in W0 by 256 and
                     'place resultant in W1 (upper 16 bits)
```

Bitwise operations use standard binary logic and 8-bit bytes:

```
B1 = %01100000
B2 = %00100010
Let B2 = B2 & B1
```

The resultant B2 will be %00100000.

Lookdown

The Lookdown command searches through a list of constants (cvalue0, cvalue1, and so on), comparing each value in the list to the search value (Svalue). If a match is found, the physical number of the term (index number) in the list is stored in the rvalue (result value) variable. Here Svalue is the search value, and cvalueX is the constant values:

```
Lookdown Svalue, (cvalue0, cvalue1, . . . cvalueN), rvalue
```

A simple example will straighten out any confusion:

```
Lookdown 5, ("16, 34, 21, 13, 7, 8, 9, 10, 5, 2"), B0
```

The command searches through the list of constants and stores the item number in B0. In this example, B0 will hold the result of 8. (Lookdown begins counting from 0 not 1.) Commas are used to delineate multiple-digit numbers.

The constant list may be a mixture of numeric and string constants. Each character in a string is treated as a separate constant with the character's ASCII value.

If the search value is not in the Lookdown list, no action is taken and the value of rvalue remains unchanged. ASCII values may be searched as well as numeric values.

```
Serin 1, N2400,B0      'Get hexadecimal character from Pin1 serially
Lookdown B0, ("0123456789ABCDEF"), B1
                       'Covert hexadecimal character in B0 to
                       'decimal value in B1.
Serout 0,N2400, (#B1)  'Send decimal value to Pin0 serially
```

Lookup

The Lookup command is used to retrieve values from a table of constants (cvalue0, cvalue1, and so on). The retrieve value is stored in the Value variable. If the index is zero, Value is set to the value of cvalue0. If the index is set to 1, then Value is set to the value of cvalue1 and so on.

```
Lookup Index, (cvalue0, cvalue1, . . . cvalueN), Value
```

If the index number is greater than the number of constants available to read, no action is taken and Value remains unchanged.

The constant may be numbers or string constants. Each character in a string is treated as a separate constant equal to the character's ASCII value. Here's an example:

```
For B0 = 0 to 5                'Set up For...Next loop
    Lookup B0, ("Hello!"), B1  'Get character number B0 from
                               'string and place in variable B1
      Serout 0,N2400, (B1)     'Send character in B1 out on Pin 0
                               'serially.
Next B0                        'Do next character.
```

Low

Low makes the specified pin an output pin and brings it low (0 volts). Only the pin number itself, 0 to 7, is specified in the command. It only works on Port B pins.

```
Low Pin
```

Here's an example:

```
Low 0      'Makes Pin0 (RB0) an output pin and brings it low
           ' ( 0 V )
```

Nap

This command places the PIC microcontroller in low-power mode for varying short periods of time. During a Nap, power consumption is reduced to a minimum.

```
Nap Period
```

The times in Table 9-4 are approximate, because the timing cycle is derived from the on-board *watchdog timer* (WDT) that is *resistor-capacitor* (RC) driven and varies from chip to chip (and over temperature).

The WDT must be enabled in the EPIC software (see EPIC software) for Nap and Sleep commands to function. If Nap and Sleep commands are not used, the WDT may be disabled. Here's an example:

```
Nap 7      'Low-power pause for 2.3 seconds
```

Output

This command makes the specified pin an output pin. Only the pin number itself, 0 to 7, is specified in the command, which works only on Port B pins. Its structure is presented here:

```
Output Pin
```

TABLE 9-4   Nap periods and delays

Period	Delay (approximate)
0	18 milliseconds
1	36 milliseconds
2	72 milliseconds
3	144 milliseconds
4	288 milliseconds
5	576 milliseconds
6	1.15 seconds
7	2.3 seconds

The following is an example:

```
Output 5     'Makes Pin5 (RB5) an output.
```

Pause

This command pauses program execution for the Period in milliseconds:

```
Pause Period
```

Period is a 16-bit number that can hold a maximum value of 65,535. In milliseconds, that works out to just over one minute (60,000 milliseconds). Unlike the other delay functions Nap and Sleep, the Pause command does not put the microcontroller into a low-power mode. This has an advantage and a disadvantage. The disadvantage is that Pause consumes more power; the advantage is that the clock is more accurate. Here's an example:

```
Pause 250     'Delay for 1/4  second
```

Peek

The Peek command reads any of the microcontroller's registers at the *Address* specified and copies the result in *Variable*. The Peek command may be used to read special registers such as A/D converters and additional I/O ports.

```
Peek Address, Variable
```

Peek reads the entire 8 bits of the register at once. If extensive bit manipulation is needed, the user may store the results of the Peek command in either B0 or B1. These two bytes may be also be used as bit variables Bit 0 to Bit 15, and extensive bit manipulation is easily performed. Byte B0 is equivalent to the range from Bit 0 to Bit 7, and byte B1 is equivalent to the range from Bit 8 to Bit 15.

The following example shows how one can check bit status. It assumes the five open pins on Port A have been configured as input pins:

```
loop:
      Peek PortA, B0                'Port A pins and copy resultant
                                    'into byte B0.
      If Bit0 = 1 then route1       'If RA0 is high jump to route1
      If Bit1 = 1 then route2       'If RA1 is high jump to route2
      If Bit2 = 1 then route3       'If RA3 is high jump to route3
      If Bit3 = 0 then route1       'If RA4 is low jump to route1
      If Bit4 = 0 then route1       'If RA5 is low jump to route2
Goto loop
```

The example shows that bits may be checked for high or low status. The Peek command also works with pins that are configured as outputs. When Peeked, the resultant shows the binary value that has been Poked in the port register. Do not use this command with the PICBasic Pro compiler.

Poke

The Poke command can write to any of the microcontroller's registers at the *Address* specified, and it copies the value in *Variable* to the register. The Poke command can also be used to write to special registers such as analog-to-digital converters and additional I/O ports.

```
Poke Address, Variable
```

Poke writes an entire byte (8 bits) to the register at once:

```
Poke 134,0       'Writes binary 0 to DDR for Port B making all pins
                 'output lines.
```

Do not use this command with PICBasic Pro compiler.

Pot

This command reads a potentiometer or another resistive transducer on the pin specified. The programmer may choose any of the Port B pins, 0 to 7, to use with this command.

```
Pot Pin, Scale, Var
```

The resistance is measured by timing the discharge of a capacitor through the resistor, usually 5K to 50K (see Figure 9-3). Scale is used to adjust the varying RC constants.

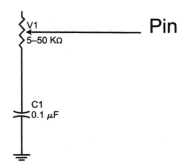

FIGURE 9-3   Reading a resistive transducer

For large RC constants, set the scale to 1. For small RC constants, set Scale to its maximum value of 255. Ideally, if the scale is set correctly, the variable Var will be set to zero at minimum resistance and to 255 at maximum resistance.

Scale must be determined experimentally. Set the device or transducer to measure at maximum resistance, and read it with a scale set to 255. Under these conditions, Var will produce an approximate value for Scale.

Many resistive-type transducers may be read using the Pot command. The important thing that separates this command from an analog-to-digital converter is that a converter measures voltage, not resistance. (Although the voltage drop across may appear similar to the Pot diagram, it is not.)

```
Pot 3,255,B0            'Read potentiometer on Pin 3 to
                        'determine scale.
Serout 0,N2400, (#B0)   'Send pot values out on pin 0
                        'serially.
```

Pulsin

```
Pulsin Pin, State, Var
```

This command measures the pulse width in 10-microsecond (µsec) increments on the Pin specified. If State is 0, the width of the low portion of the pulse is measured. If State is 1, the width of the high portion of the pulse is measured. The measured width is stored in variable Var. The variable Var is a 16-bit number and therefore can contain numbers from 0 to 65,535:

To calculate the measured pulse width, multiply Var by 10 µsec:

```
Var * 10 µsec = measured pulse width
```

Pulse widths from 10 to 655,350 µsec can be measured.

If the pulse width is larger than the maximum width, the microcontroller can measure Var as set to zero. If an 8-bit variable is used for Var, only the lower byte (LSB) of the 16-bit measurement is stored. The command may use any Port B pin from 0 to 7.

```
Pulsin 2,0,W2           'Measures low pulse on Pin 2 (RB2) and
                        'places width measurement * 10 µsec in
                        'W2
```

Pulsout

This command generates a pulse on the Pin specified. The pulse width is specified by Period. The variable Period is a 16-bit number that can range from 0 to 65,535:

```
Pulsout Pin, Period
```

The pulse width is calculated by multiplying the variable Period by 10 μsec:

```
Period * 10 μsec = pulse width
```

Therefore, pulse widths from 10 to 655,350 μsec may be generated.

Pulses are generated by toggling the pin twice. Thus, the initial state of the pin, 0 or 1, determines the polarity of the pulse.

If the initial state of the pin is low, Pulsout outputs a positive pulse. On the other hand, if the initial state of the pin is high (+5 volts), Pulsout outputs a negative (0 volts) pulse.

The Pulsout command may use any Port B pin from 0 to 7. The pin used is automatically made into an output pin.

```
Low 6                 'Set pin6 (RB6) to an output and bring it
                      'low
Pulsout 6,1000        'Send a positive pulse 10,000 μsec (10
                      'millisecond) long out on Pin 6 (RB6).
```

PWM

This command outputs a *pulse width modulation* (PWM) train on the pin specified. Each cycle of Pwm consists of 256 steps. The Duty cycle for each Pwm ranges from 0 (0 percent) to 255 (100 percent). This Pwm cycle is repeated according to the number specified by Cycle. The Pwm command may use any Port B pin from 0 to 7.

```
Pwm Pin, Duty, Cycle
```

The pin becomes an output just prior to pulse generation and reverts to an input after generation stops. This allows a simple RC circuit to be used as a simple digital-to-analog converter (see Figure 9-4).

```
Pwm 7,126,155         'Send a 50% duty cycle PWM signal out on
                      'Pin 7 (RB7) for 155 cycles.
```

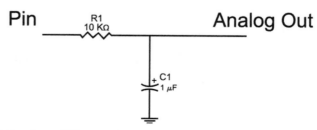

FIGURE 9-4   The RC circuit as a D/A converter

NOTE: If the Pwm command is used to control a high-current device, the output signal should be buffered.

Random

This command generates a pseudo-random number in Var. The variable *Var* must be a 16-bit variable. Random numbers range from 1 to 65,635 (zero is not produced).

```
Random Var
```

Here's an example:

```
Random W2      'Generate random number in W2
```

Read

This command reads the on-chip EEPROM (if available) at the specified Address, and the resultant byte at the address is copied in the Var variable.

```
Read Address, Var
```

If Address is 255, Var returns with the number of EEPROM bytes available. This instruction may only be used with microcontrollers that contain on-chip EEPROM, such as the 16F84. Here's an example:

```
Read 5, B0     'Read EEPROM location number 5 and copy
               'into B0.
```

Return

This command causes program execution to return from a called Gosub command, as shown here:

```
Gosub send1                      'Jump to subroutine labeled send1
. . .                            'Program returns here
. . .
. . .
send1:                           'Subroutine send1 begins
Serout 0,N2400, ("Hello!")       'Send "Hello!" out on pin0 serially
Return                           'Return to main program
```

Reverse

This command reverses the status of the Pin specified. If Pin is an output, it is reversed into an input, and vice versa. Only the pin number itself, 0 to 7, is specified in the command. Reverse works only on Port B pins. Its structure is as follows:

```
Reverse Pin
```

Here's an example:

```
Output 3      'Makes Pin 3 (RB3) an output pin
Reverse 3     'Change Pin 3 (RB3) to an input pin
```

Serin

This command enables the microcontroller to receive serial data on the Pin specified. The data is received in standard asynchronous mode using eight data bits, no parity, and one stop bit (see Figure 9-5). Mode sets the baud rate and TTL polarity as shown in Table 9-5. Serin's structure is as follows:

```
Serin Pin, Mode, [(Qual {,Qual}),] Item {,Item}
```

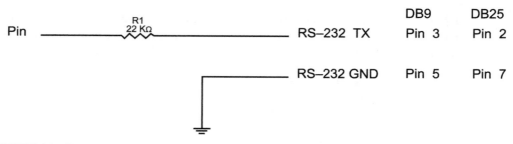

FIGURE 9-5 Pin setup using Serin

TABLE 9-5   Serin baud rate and TTL polarity

Symbol	Baud rate	Polarity
T2400	2400	TTL True
T1200	1200	TTL True
T9600	9600	TTL True
T300	300	TTL True
N2400	2400	TTL Inverted
N1200	1200	TTL Inverted
N9600	9600	TTL Inverted
N300	300	TTL Inverted

Here's an example of Serin at work:

```
'Convert decimal number to hexadecimal
Loop:
Serin 1, N2400, B0                    'Receive decimal
                                      'number on Pin1, 2400
                                      'Baud store in B0
Lookup B0, ("0123456789ABCDEF"), B   'Use B0 as index
                                      'number and lookup
                                      'hex equivalent.
Serout 0, N2400, (B1, 13, 10)        'Transmit hex
                                      'equivalent out on Pin0    'serially with
carriage
                                      'return (13) and line
                                      'feed (10)
Goto Loop                            'Do it again
Triggers
```

The microcontroller can be configured to ignore all serial data until a particular byte or sequence of bytes is received first. These bytes are called qualifiers and are enclosed within parenthesizes. Serin must receive these bytes in their exact order (if more than one) before receiving data. If a byte does not match the next byte in a qualifying sequence, the qualification process resets. If this happens, the next byte received is compared to the first item in the qualification sequence. Once the qualification is met, Serin begins receiving data.

The qualifier can be a constant, variable, or string. Each character of a string is treated as an individual qualifier. In the following line, Serin waits until the character A is received serially on Pin1 and then puts the next character in B0:

```
Serin 1, N2400, ("A"), B0
```

Serout

This command enables the microcontroller to transmit serial data on the Pin specified. The data is transmitted in standard asynchronous mode using eight data bits, no parity, and one stop bit (see Figure 9-6). Mode sets the baud rate and TTL polarity as shown in Table 9-6. Its structure is as follows:

```
Serout Pin, Mode, Item {,Item}
```

Serout supports three types of data that may be mixed and matched freely within a single Serout statement:

- A string constant, which is transmitted as a literal string of characters.

- A numeric value (either a variable or a constant), which will transmit the corresponding ASCII character. It is used often to transmit a carriage return (13) and a line feed (10).

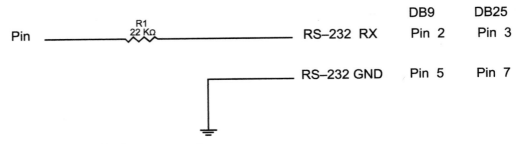

FIGURE 9-6 Using the Serout command

TABLE 9-6 Serout baud rate and TTL polarity

Symbol	Baud rate	Polarity
T2400	2400	TTL True
T1200	1200	TTL True
T9600	9600	TTL True
T300	300	TTL True
N2400	2400	TTL Inverted
N1200	1200	TTL Inverted
N9600	9600	TTL Inverted
N300	300	TTL Inverted
OT2400	2400	Open Drain
OT1200	1200	Open Drain
OT9600	9600	Open Drain
OT300	300	Open Drain
ON2400	2400	Open Source
ON1200	1200	Open Source
ON9600	9600	Open Source
ON300	300	Open Source

- A numeric value proceeded by a pound sign (#), which will transmit an ASCII representation of its decimal value. For instance, if W0 equals 123, then #W0 (or #123) will transmit as 1, 2, or 3.

```
Serout 0,N2400, (#B0,10)        'Send the ASCII value of B0 followed by
                                'a linefeed out Pin 0 serially.
```

NOTE: Single-chip RS-232-level converters are common and inexpensive (MAXIM's MAX232), and they should be implemented when needed or to insure proper RS-232 communication.

Sleep

This command places the microcontroller in low-power mode for the Period specified in seconds:

```
Sleep Period
```

Since Period is a 16-bit number, delays up to 65,535 seconds (a little over 18 hours) are possible. Sleep uses the WDT on the microcontroller that has a resolution of 2.3 seconds (refer to the Nap command). Here's an example:

```
Sleep 120     'Sleep (low power mode) for two minutes.
```

Additional Sleep Notes

It has been determined that Sleep may not work properly on all PICmicros. During Sleep calibration, the PICmicro is reset. Different devices respond in different ways to this reset. For instance, many registers may be altered, notably the TRIS registers that set all the PORT pins to inputs.

However, the TRIS register for Port B is automatically saved and restored by the Sleep routine. Any other Port directions must be reset by the user program after Sleep. Other registers may also be affected. See the microconntroller individual data sheets for more information.

To get around potential problems, an uncalibrated version of Sleep has been added. This version does not cause a device reset so it has no effect on any of the internal registers. All the registers, including Port direction, remain unchanged during and after a Sleep instruction.

However, actual Sleep times will no longer be as accurate and will vary depending on device particulars and temperature. To enable the uncalibrated version of Sleep, add the following lines to a PICBasic Compiler program:

```
asm
SLEEPUNCAL = 1
endasm
```

The PICBasic compiler software is packaged with a PICmicro macro assembler (PM.exe). Although you will not write any assembly code, it is available to those who have familiarity with assembly language and PBC library routines.

Sound

This command generates tones and/or white noise on the specified Pin. Note 0 is silence. Notes 1 through 127 are tones, and notes 128 through 255 are white noise. Tones and white noises are in ascending order. Duration is a numeric variable from 0 to 255 that determines how long the specified note is played. Each increment in duration is equivalent to approximately 12 milliseconds. The Sound command may use any Port B pin from 0 to 7.

```
Sound Pin, (Note, Duration {, Note, Duration})
```

The waveform output is TTL-level square waves. A small speaker and capacitor can be driven directly from the microcontroller pin (see Figure 9-7). Piezo speakers may be driven directly. Here's an example:

```
Sound 4, (100,10,50,10)    'Plays 2 notes consecutively on Pin 4
                           '(RB4).
```

Toggle

This command inverts the state of the specified Pin. The pin specified is automatically made into an output pin. Toggle may use any Port B pin from 0 to 7.

```
Toggle Pin
```

Here's an example:

```
High 1      'Make Pin1 (RB1) high
Toggle 1    'Invert state of Pin 1 and bring it low
```

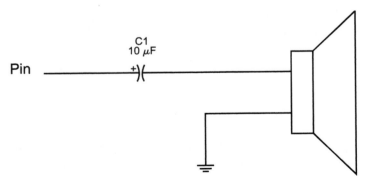

FIGURE 9-7   Capacitor and speaker run by the microcontroller pin

Write

This command writes the Value to the on-chip EEPROM (if available) at the specified Address:

```
Write Address, Value
```

This instruction may only be used with microcontrollers that contain on-chip EEPROM such as the 16F84. An example is provided here:

```
Write 5,B0      ' Write the value in B0 to EEPROM address 5
```

Additional Command Reference
for PICBasic Pro

In this chapter, you will review the additional commands available to the PICBasic Pro compiler (see Table 10-1). These commands, as well as the commands outlined in the previous chapter (Chapter 9, "PICBasic Language Reference"), comprise the full vocabulary available to the PICBasic Pro user.

TABLE 10-1   Additional PICBasic Pro commands

@	Inserts one line of assembly code.
ADCIN	Reads on-chip analog-to-digital converter.
ASM . . . ENDASM	Inserts assembly language code section.
BRANCHL	Branches out of page (long Branch).
Clear	Zeroes all variables.
Clearwdt	Clears *Watchdog Timer* (WDT).
Count	Counts number of pulses on a pin.
Data	Defines initial contents of on-chip *electrically erasable programmable read-only memory* (EEPROM).
Debugin	Asynchronizes serial input.
Disable	Disables "On Interrupt" and "On Debug" processing.
Disable Debug	Disables "On Debug" processing.
Disable Interrupt	Disables "On Interrupt" processing.
DTMFOUT	Produces *dual-tone multifrequency* (DTMF) (phone) tones on a pin.
Enable	Enables "On Interrupt" and "On Debug" processing.
Enable Debug	Enables "On Debug" processing.
Enable Interrupt	Enables "On Interrupt" processing.
Freqout	Produces up to two frequencies on a pin.
Hserin	Hardware asynchronous serial input.
Hserout	Hardware asynchronous serial output.
I2cread	Reads bytes from I2C device.
I2cwrite	Writes bytes to I2C device.
If . . . Then	Performs a comparison test and related functions.
Lcdin	Reads from *liquid crystal display* (LCD) *random access memory* (RAM).

(continued)

TABLE 10-1   Additional PICBasic Pro commands *(continued)*

Lcdout	Displays characters on LCD.
Lookdown2	Searches constant/variable table for value.
Lookup2	Fetches constant/variable from table.
On Debug	Executes Basic debug monitor.
On Interrupt	Executes Basic subroutine on an interrupt.
Pausesus	Delays with 1-microsecond (usec) resolution.
Rctime	Measures pulse width on a pin.
Readcode	Reads word from code memory.
Resume	Continues execution after interrupt handling.
Serin2	Asynchronous serial input (BS2 style).
Serout2	Asynchronous serial output (BS2 style).
Shiftin	Asynchronous serial input.
Shiftout	Asynchronous serial output.
Stop	Stops program execution.
Swap	Exchanges the values of two variables.
While . . . Wend	Executes statements while condition is true.
Writecode	Writes word to code memory.
Xin	X-10 input.
Xout	X-10 output.

@

This symbol, when used at the beginning of a line, allows you to insert one assembly language statement into your PICBasic Pro program.

```
@ Assembly Statement
```

A simple example follows:

```
B0 var byte
@     rlf     B0,F
```

The first statement defines the variable B0. The second statement inserts an assembly language command that shifts the bits in the variable B0 one position to the left.

Although this is not a book on assembly language coding, it is nice to be able to throw some assembly commands into the basic programs. To really see how this works, let's first construct a little circuit (see Figure 10-1).

What the following PICBasic Pro program accomplishes is to light the four *light-emitting diodes* (LEDs) illustrated in Figure 10-1 sequentially:

```
'PICBasic Pro Program 10.1
'Initialize Port B
TRISB = 0
PORTB = 1
Pause 250
Loop:
PORTB = PORTB * 2
IF PORTB > 16 Then
```

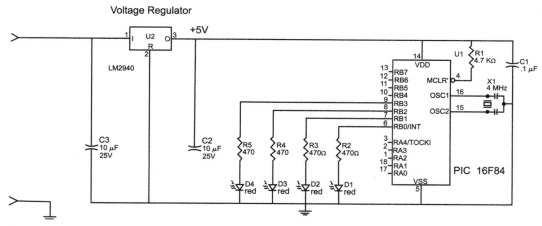

FIGURE 10-1 The circuit with four LEDs

```
PORTB = 1
EndIF
Pause 250
GoTo Loop
'End
```

The same effect can be accomplished by removing a few basic lines and substituting some *machine-language* (ML) code. Here's the revised program:

```
'PICBasic Pro Program 10.2
'Initialize Port B
TRISB = 0
PORTB = 1
Loop:
@ rlf PORTB,F
Pause 250
IF PORTB > 16 Then
PORTB = 1
EndIF
GoTo Loop
'End
```

This program has a small bug. Sometimes when a circuit is turned on, the carry bit in the status register is set. When the assembly line "@ rlf PORTB,F" rotates the bits, the carry bit is shifted into the LED display. This shows as two lit LEDs instead of one. I am not going to discuss assembly code here, except to say that you can correct this problem by clearing the carry bit with a second assembly line command, as shown in the program here:

```
'PICBasic Pro Program 10.3
'Initialize Port B
```

```
TRISB = 0
PORTB = 1
Loop:
IF PORTB > 16 Then
PORTB = 1
EndIF
Pause 250
@ bcf 3,0
@ rlf PORTB,F
GoTo Loop
'End
```

Now the program functions perfectly. You may wonder what would be the advantage of using assembly code inside a PICBasic Pro program. There are two advantages: speed and size.

Any assembly code (or routine) inserted inside a PICBasic Pro program will execute at lightning-fast speed. Size is the second consideration. Assembly code takes up less room. For instance, the PICBasic Pro program size is 86 words as compared to the two programs with the inserted ML code that have a size of 55 and 56 words respectively.

Adcin

This command reads the on-chip analog-to-digital converter. The particular PIC microcontroller you are programming must of course be equipped with a built-in analog-to-digital converter.

```
Adcin Channel, Var
```

Asm . . . EndAsm

These instructions allow you to add multiple assembly language lines into a PICBasic program. This is in contrast to the @ command that just allows a single line of assembly code to be inserted at a time. Its structure is as follows:

```
ASM
EndAsm
```

If you rewrite your LED sequential lighting program using these commands, the program would look like this:

```
'PICBasic Pro Program 10.4
'Initialize Port(s)
TRISB = 0
PORTB = 1
Loop:
IF PORTB > 16 Then
PORTB = 1
EndIF
Pause 250
```

```
'Begin Assembly language routine
Asm
bcf 3,0
rlf PORTB,F
EndAsm
'End Assembly Language routine
GoTo Loop
'end
```

Branchl

Branchl (Branch Long) works in the same manner as Branch (refer to Branch in Chapter 9). The difference is that the Branchl command can index past the one-code-page (2K) limit of the Branch command.

```
Branchl     Index, [Label1 {, Label ...}]
```

Clear

This command sets all RAM registers and all variables to zero.

Clearwdt

This command clears the WDT, which is a free-running on-chip RC oscillator that does not require any external components. It is used in conjunction with either the Sleep or Nap commands.

Count

This command counts the number of pulses that occur on Pin during the Period and stores the results in Var. The resolution for Period is in milliseconds.

```
Count Pin,Period,Var
```

The pin identified in the Count command is automatically made into an input pin. The command checks the state of the pin and counts every low to high transition. With a 4 MHz crystal, the pin is polled every 20 µsec. With a 20 MHz crystal, the pin is checked every 4 µsec.

Here's an example:

```
'Count the number of pulses on pin RB2 in 1 second.
Count PortB.2, 1000, W1
```

Data

The Data command stores constants in the PICs nonvolatile EEPROM when the device is programmed, not each time the program is run. Location specifies the starting location for the storage. If omitted, storage begins at address 0. The structure is as follows:

```
Data {@Location,} Constant {,Constant...}
```

Here's an example of the command at work:

```
'Store numbers 7, 11 and 14 at starting at location 3
Data @3,7,11,14
```

Debug

The Debug command sends asynchronous serial information (an item) to a predefined pin at a predefined baud rate: a data format with 8 bits, no parity, and 1 stop bit (8N1). Debug is typically used to send program-debugging information, such as variables, to a terminal program. Its structure is as follows:

```
Debug     Item {,Item...}
```

The following statements define the serial pin and baud rate for debug:

```
'Set Debug pin port
Define Debug_Reg PORTB

' Set Debug pin bit
Define Debug_Pin Bit 0

'Set Debug baud rate
Define Debug_Baud 2400

'Set Debug mode: 1 = inverted, 0 = true
Define Debug_Mode 1
```

Debugin

```
Debugin {Timeout, Label,}[Item{,Item...}]
```

Debugin receives asynchronous serial information (an item) to a predefined pin at a predefined baud rate: a data format with 8 bits, no parity, and 1 stop bit (8N1).

Timeout and Label are options you may include that enable the program to continue if a particular character (number) is not received within the allocated time. Timeout is specified in increments of 1 millisecond.

The following statements define the serial pin and baud rate for Debugin:

```
'Set Debugin pin port
Define Debugin_Reg PORTB

' Set Debugin pin bit
Define Debugin_Pin Bit 0
```

```
'Set Debugin baud rate
Define Debugin_Baud 2400

'Set Debug mode: 1 = inverted, 0 = true
Define Debgin_Mode 1
```

Disable

The Disable command disables both the debug and interrupt processing. To enable these processes, one must use the Enable instruction.

Disable Debug

Disable Debug disables the debug processing. To enable debug, one must use the Enable Debug instruction.

Disable Interrupt

Disable Interrupt disables the interrupt processing. To enable interrupt processing, one must use the Enable Interrupt instruction.

DTMFout

This command enables you to output standard telephone (DTMF) Touch-Tones. Pin is the specified output pin, onms (on milliseconds) is the number of milliseconds to output the DTMF tone, and offms (off milliseconds) is the number of milliseconds to pause between each tone. If onms is not specified, it defaults to 200 ms. If offms is not specified, it defaults to 50 milliseconds.

```
Dtmfout Pin, {onms, offms,} [Tone {,Tone...}]
```

The Tone parameter is numbered 0 to 15. The tones from 0 to 9 are the same tones as you find on a telephone keypad. Tone number 10 is the star (*) key and Tone 11 is the pound (#) key. Tones 12 to 15 correspond to the extended keys A through D.

microEngineering Labs Inc. suggests using the following filter (see Figure 10-2) when using DTMF tones. They also recommend using a faster than standard 4.0 MHz crystal oscillator.

```
'send DTMF tones for area code 718 on port A pin 1
Dtmfout porta.1, [7,1,8]
```

Enable

Enable is an instruction that enables debug and interrupt processing previously disabled with the Disable instruction.

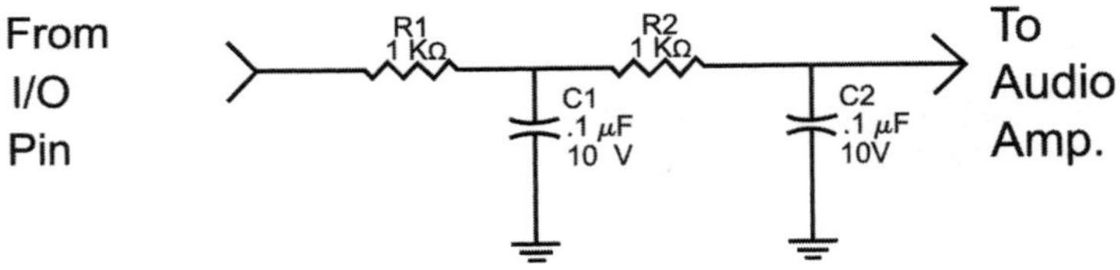

FIGURE 10-2   The recommended filter

Enable Debug

Enable Debug enables the debug processing that was previously disabled.

Enable Interrput

Enable Interrupt enables the interrupt processing that was previously disabled.

Freqout

This command produces the frequency on the specified pin for the specified amount of time (onms, milliseconds) in milliseconds. Up to two frequencies from 0 to 32,767 Hz may be produced at one time.

```
Ffrqout Pin, Onms, Frequency1 {, frequency2}
```

As with the DTMFOUT command, microEngineering Labs recommend using a faster than standard (4.0 MHz) oscillator preferably 20 MHz in addition to adding the following filter to the output pin to smooth the sine wave and remove some of the harmonics:

```
'Output a 2 kHz tone on pin3 for 3 seconds
Freqout PORTB.3, 3000,2000
```

Hserin

Hserin is a serial function that utilizes a hardware serial port (USART) on the PIC microcontroller. The serial parameters and baud rate are specified using Defines.

```
Hserin {ParityLabel,} {Timeout,Label,} [Item{,...}]
```

Hserout

Hserout is a serial function that utilizes a hardware serial port (USART) on the PIC microcontroller. The serial parameters and baud rate are specified using Defines.

```
Hserout [Item{,Item...}]
```

I2cread

I2creaed is used to read data from a serial EEPROM using a two-wire I2C interface (it is similar to PICBasic I2cin).

```
I2cread DataPin, Clockpin, Control, {Address,} [Var {,Var...}] {,Label}
```

I2cwrite

I2cwrite is used to write data to a serial EEPROM using a two-wire I2C interface (it is similar to PICBasic I2cout).

```
I2cwrite DataPin, ClockPin, Control, {Address,} [Value{,Value...}] {,Label}
```

If . . . Then

The PICBasic Pro version of this command is much richer than the standard PICBasic command. It performs a comparison test and can function in two modes as shown here:

```
If Comp {and/or Comp} Then Label
If Comp {and/or Comp} Then
     Statement . . .
Else
     Statement  . . .
EndIF
```

In the first mode, if the particular condition is met (true), then the program execution jumps to the statements beginning at the Label. If the condition is not true, program execution continues at the next line. In this first mode, the Then in the If . . . Then is essentially a Goto.

The command compares variables to constants or to other variables. If only one variable is used in a comparison, it must be placed on the left. All comparisons are unsigned.

The following is a list of valid comparisons:

- =   Equal
- \>   Greater than
- <   Less than
- <>   Not equal
- <=   Less than or equal to
- >=   Greater than or equal to

If the value in variable B8 is less than or equal to 25, then the program jumps to loop:

```
If B8 <= 25 then loop
```

The If . . . Then command can also be used with two binary logic comparisons, And and Or.

In the second mode, the If . . . Then can conditionally execute a group of statements following the Then. These statements must be followed by an Else or Endif to complete the command:

```
If B8 <> 25 Then
      B8 = 100
      B7 = 25
Endif
```

or

```
If B8 = 25 Then
      B7 = 25
Else
      B7 = 10
Endif
```

Lcdin

Read the RAM present in a LCD and store the data to variable Var.

```
Lcdin {Address,} [Var {Var ...}]
```

Lcdout

This command displays items on an LCD that uses a Hitachi 44780 controller or equivalent. Many LCD modules have either a 14- or 16-pin header. The LCD controller supported by the PBP compiler is common.

```
Lcdout Item {,Item...}
```

The following schematic in Figure 10-3 shows how to connect an LCD module to a 16F84. Once the LCD module is properly connected to the microcontroller, data sent to it will appear on the display. For example, if you send "Hello," then "Hello" appears on the display. The cursor (printing position) automatically moves from left to right:

```
Lcdout     "Hello World"     'Prints 'Hello World' on LCD display
```

You can also send instructions to the LCD module. To identify a particular byte as an instruction, precede it with the instruction prefix character ASCII 254 (0xFE hex, 11111110 binary). The interface treats the byte immediately after the prefix as an instruction and then automatically returns to data mode.

As an example, the clear-screen instruction is ASCII 1. To clear the screen, send <254><1> (where the < and > symbols mean single bytes set to these values, not text as typed from the keyboard). See Table 10-2.

```
Lcdout 254,1, "Hello World"     'Clears display and prints "Hello World"
```

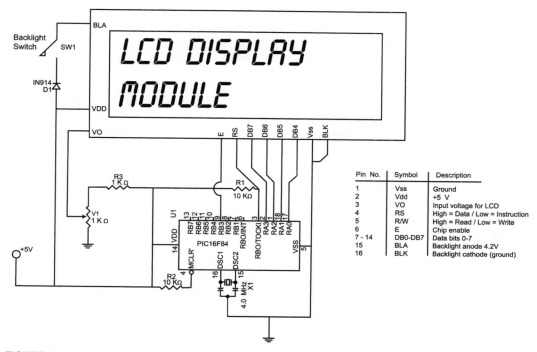

FIGURE 10-3   Schematic of the LCD connected to a 16F84 microcontroller

TABLE 10-2   LCD instruction codes

Instruction	Code (decimal)
Clear screen	1
Home position (move cursor top left of display)	2
Move cursor one character position left	16
Move cursor one character position right	20
Scroll display one character position left	24
Scroll display one character position right	28
Set cursor position (*double data random access memory* [DDRAM] address)	128 + addr
Set point in *character generator* (CG) RAM	64 + addr

Lookdown2

The Lookdown2 command searches through a list of values for the search value. If it is found, the index of the matching value is stored in Var. If the value found is the first item in the list, Var will contain a zero. If second in the list, Var will contain a value of 1. If no match is found, the value in Var remains unchanged.

```
Lookdown2 Search, {Test} [Value {,Value...}],Var
```

This function is similar to the standard Lookdown command compatible to both PICBasic and PICBasic Pro compilers. The differences are thus: Lookdown2 can handle 16-bit values, and there is also the optional {Test} parameter. The {Test} parameter can be set to perform other comparisons besides the standard equal to (=). For instance, {Test} could be set to test for the first value less than (<) the search value.

Lookdown2 generates code that is approximately three times larger than the standard Lookdown command. If you only need to search 8-bit values, you may want to use the standard Lookdown command.

Lookup2

The Lookup2 statement is used to retrieve data from a list of Values. If the Index is set to 0, the first value in the list is stored in Var. If the Index is set to 1, the second value in the table is stored in Var and so on. If the number in Index is greater than or equal to the number of entries in the list, no action will be taken and Var remains unchanged.

```
Lookup2    Index, [Value{,Value. . .}],Var
```

This function is similar to the standard Lookup statement compatible to both PICBasic and PICBasic Pro compilers. The differences are that Lookup2 can handle 16-bit values.

Lookup2 generates code that is approximately three times larger than the standard Lookup command. If you only need to search 8-bit values, you may want to use the standard Lookup command.

On Debug

On Debug permits a debug monitor to be executed between every PICBasic Pro instruction.

```
On Debug Goto Label
```

On Interrupt

On Interrupt enables the handling of interrupts by a subroutine. Interrupts are an important programming feature, so I will spend a little time on this command to explain how to use it. On the PIC 16F84A, four sources of interrupt are possible: an external signal on pin RB0/INT, a TMR0 overflow, a PORTB change (RB7: RB4), and a data EEPROM write complete.

```
On Interrupt Goto Label
```

When an interrupt occurs, it sets a flag (bit) in a register. The register is one byte of memory at address (0Bh, 8Bh), which is called the interrupt control register (INTCON). This register records individual interrupt requests by setting individual bits that correspond to the particular interrupt. For instance, an interrupt (external signal) on pin RB0/INT will set bit 1 in the INTCON register. The bits, when used in this type of signaling application, are called flags. The microcontroller responds to this flag being set by running a interrupt routine.

In addition to recording interrupts, the INTCON register uses other bits to either enable or disable interrupts. The INTCON register also contains the global interrupt enable bit GIE (bit 7).

For any interrupt to be recognized, the GIE bit must be enabled (set). If the GIE bit is clear (0), no interrupt will be responded to. In addition, the individual interrupts may also be enabled or disabled by setting or clearing their control bit in the INTCON register. For instance, the enable bit for RB0/INT interrupt is bit 4.

To enable interrupts on RB0/INT, you must set the GIE bit (bit 7) as well as the RB0/INT enable bit (bit 4). You must enable the interrupt you want to capture before using the On Interrupt command.

In PICBasic Pro, you can enable interrupts to RB0/INT with the following command:

```
INTCON = %1001000
```

or

```
INTCON = 144
```

When an interrupt is signaled, the microcontroller does not immediately jump to the interrupt subroutine called out in the command Label. It must first complete whatever command it is currently processing. This can create a delay from when an interrupt is signaled to when it is actually handled. In some cases, the delay can be considerable. Consider the following command:

```
Pause 10000      'pause 10 seconds
```

If an interrupt is signaled during the execution of this command, there could be close to a 10-second delay before the interrupt is responded to. This is far too long for any time-critical response. So when using interrupt processing in your programming, avoid using a delay like that. Instead, rewrite the delay like this:

```
For x = 1 to 10000
pause 1
next x
```

Using this style delay, the interrupt response time is only a few milliseconds.

Two more statements need to be discussed before we can write a simple interrupt program, Enable and Disable.

The Disable statement allows sections of PICBasic programs to be executed without the possibility of being interrupted. The Disable statement is placed just before our interrupt subroutine. The Enable statement restores the microcontroller's ability to respond to interrupts.

The Disable and Enable statements are not really commands. These are compiler directives. When the compiler encounters the Disable statement in a program, it codes the preceding PICBasic program not to respond to interrupts. The compiler continues to write noninterruptible code until it encounters the Enable statement. This is the reason that the Disable statement is placed just before our interrupt handler subroutine and the Enable statement is placed right after. Here's a simple interrupt program:

```
'PICBasic Pro program 10.5
'This program responds to an interrupt on pin RB0
On Interrupt Goto Light
INTCON = 144      'Set up Interrupt Control Register to enable RB0/INT
trisB = 1         'Set up TRISB register
portb.1 = 0       'Turn off LED
'
start:
'                 Main Program Goes Here
pause 1
goto start
'
'
'                 Interrupt Routine follows
\'
Disable           'Disable all interrupts
Light:'Light an LED connected to RB1
PortB.1 = 1       'Turn on LED
Resume            'Return to main program
Enable            'Enable interrupts
```

This is a very simple program; it stays in a loop. An interrupt on RB0 will cause an LED connected to PortB.1 to light. To trigger an interrupt, place a signal on pin RB0. Figure 10-4 places a signal on RB0 using a pull-up resistor and switch. The signal is just raising the voltage and lowering the voltage on pin RB0 to +5 volts and then to ground.

The Option_Reg register (Bit 6) allows you to choose if the interrupt on RB0 will be generated on a low- to high-transition or a high- to low-transition signal. See the data sheets on the 16F84A for more information.

Pauseus

Pauseus is similar to pause with the exception that the increments are in microseconds. Its structure is as follows:

```
Pauseus Period
```

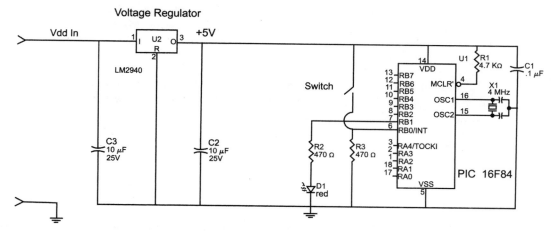

FIGURE 10-4   Placing a signal on RB0 using a pull-up resistor and switch

The variable Period is a 16-bit value so it can hold a number up to 65,535 (microseconds). Here's an example of the command:

```
Pauseus 1000       'Delay 1 millisecond
```

Peek

Same as the command in PICBasic, Peek should not be used in PICBasic Pro programs.

```
Peek Address,Var
```

Poke

Same as the command in PICBasic, Poke should not be used in PICBasic Pro programs.

```
Poke Address,Var
```

Pulsin

This command measures the pulse width on a pin. The State option determines if the high or low portion of the pulse width is measured. If State equals 1, it is the high portion being measured. If State equals 0, it is the low portion being measured. The measurement of the pulse is stored in the variable Var. If the pulse edge never happens or the width is too great, Var is set to 0. If the variable used (Var) is only an 8-bit (byte) variable, only the LSB of the 16-bit measurement is stored in Var.

```
Pulsin Pin,State,Var
```

Resolution using a 4.0 MHz Xtal is in 10 µsec increments. If a 20 MHz Xtal is used, the resolution will be 2 µsec. Defining an OSC value has no effect on the Pulsin command.

```
'Measure high pulse on RB1
X var word
Pulsin PORTB.1,1,X
```

Pulsout

Pulsout generates a pulse of Pin for the specified Period. Pulses are generated by toggling the pin twice. The initial pin state determines the polarity of the pulse. A pin that is high (1) in its initial state will generate a negative pulse, and a pin that is initially low will generate a positive pulse.

```
Pulsout Pin, Period
```

The resolution of the pulse is determined by the oscillator used on the microcontroller. If the standard 4.0 MHz Xtal is used, the period is in 10 µsec increments. At 20 MHz, the resolution increases to 2 µsec per increment. Defining an OSC value has no effect on the Pulsout command, and Pin identifies which pin on the microcontroller is to be used.

```
'Send a 2 millisecond pulse (4MHz oscillator) to PIN 1
Pulsout PortB.1, 200
```

RCtime

RCtime is used to measure resistive devices, typically between 5 and 50 kilo-ohms. Resistance is taken by measuring the charge or discharge of a capacitor through the resistive device to be measured. Pin identifies which pin on the microcontroller is to be used and State determines which option is used; state 0 equals a charge, and state 1 equals a discharge. Var is the variable that holds the results of the command.

```
RCtime Pin, State, Var
```

If the Pin doesn't change state, 0 is returned. Here's an example:

```
B0 var word            'initialize variable
Low PortB.1            'discharge capacitor
pause 10               'discharge for 10 milliseconds
RCtime PortB.1 ,0, B0  'read resistance on pin1
```

Readcode

This command reads word-sized code at the specified address and stores the data in Var:

```
Readcode Address, Var
```

Here's an example:

```
B0 var word
Read 100, B0        'Copy value at Address 100 into B0
```

Resume

This command sends program execution back to where it left off after handling an interrupt. If the optional Label is used, program execution will continue at the Label instead of where it was when it was interrupted. See On Interrupt for an example. The Resume structure is as follows:

```
RESUME {Label}
```

Serin2

This is an advanced form of a standard serial using the Serin command. The most notable features are FlowPin, Timeout, and Label.

```
Serin2 Datapin {\Flowpin}, Mode, {ParityLabel,}{Timeout, Label,} [Item...]
```

Flowpin is the optional flow control pin that may be assigned to prevent data from overrunning the receiver. An optional timeout may be included in the program so that if a serial byte is not received in a specified amount of time the program continues at the Label. Timeout is specified in units of 1 millisecond.

Serout2

Serout2 is an advanced form of the standard serial out Serout command. The most notable features are FlowPin, Timeout, and Label.

```
Serout2 DataPin {\FlowPin},Mode, {Pace,}{Timeout,Label,}[Item...]
```

Here Flowpin is the optional flow control pin that may be assigned to prevent data from overrunning the receiver. An optional timeout may be included into the program so that if the Flowpin does not change into its enabled state the specified amount of time the program continues at the Label. Timeout is specified in units of 1 millisecond.

Shiftin

This is a synchronous serial in command that uses ClockPin to shift bits in on the DataPin. Information is stored in variable Var.

```
Shiftin DataPin, ClockPin, Mode, [Var{\bits}...]
```

Shiftout

Shiftout is a synchronous serial out command that uses ClockPin to shift bits in on DataPin. Information is stored in Variable Var.

```
Shiftout DataPin, ClockPin, Mode, [Var{\Bits}...]
```

Swap

Swap exchanges the values between two variables. The command may be used with bit, byte, and word variables.

```
Swap    Variable, Variable
```

Here's an example:

```
Swap B0, B1
```

While...Wend

```
While Condition
     Statement ...
Wend
```

This repeats execute statements when the While condition is true. When the condition is no longer true, conditions are executed at the statement following the Wend. Condition is also a comparison expression.

```
D0 = 1
While D0 <=100
     PortB.1 = 1
     D0 = D0 + 1
Wend
     PortB.1 = 0
```

Writecode

This command writes a word-sized variable into memory at the location Address. It can be used to write self-modifying code.

```
Writecode Address,Value
```

Xin

Xin receives X-10 data and stores the housecode and keycode in Var. X-10 system and plug-in controllers are used for home and office automation. The X-10 system uses a home's (or office's) existing electrical system to transmit information to the X-10 modules. X-10 modules are plugged into electrical wall outlets. The light or electrical appliance you want to control is then plugged into the X-10 module, which controls the current to that electrical device or appliance. Each X-10 module is given a unique identification (housecode and keycode) that separates it from the other X-10 modules in the system. To turn a device on or off, the X-10 codes are transmitted though the electrical system, and all the X-10 modules are constantly listening to the power line. If they receive an X-10 transmission that matches their housecode and keycode, it will respond to the transmission's command. The XIN structure is as follows:

```
Xin DataPin, ZeroPin, {Timeout, Label,} [Var{,...}]
```

The X-10 transmits its coded signal on the AC power line, when the line is most quiet at zero crossing. To understand this, visualize a sine wave, which represents the AC power in your home. The point where the sine wave passes through the imaginary horizontal line at zero volts is the zero crossing. In other words, it is where the voltage travels from the positive side of the sine wave to the negative side and vice versa.

Playing around with this AC power can be dangerous. To make the job safer, X-10 developed a product called the TW523 Power Line interface. The TW523 isolates the power line from the user and allows the use of *Time to Live* (TTL) logic signals. The TW523 has a TTL output signal to inform the microcontroller of zero crossing. It has a receive line for the microcontroller to receive X-10 code signals and an output line for the microcontroller to transmit X-10 codes.

Xout

Xout transmits housecode followed by keycode, and it repeats this a certain number of times. Housecode is a number between 0 and 15 that corresponds to the housecode set on the X-10 module A through P. Keycode is a number between 0 and 15 corresponding to the keycode set on the X-10 Module 1 through 16. DataPin is automatically made into an outline pin and transmits X-10 coded data.

ZeroPin is automatically made into a input pin. It waits for the zero-crossing data from the X-10 module to signal the transmission of X-10 data. ZeroPin is pulled up to 5 volts using a 4.7K resistor.

```
Xout DataPin, ZeroPin, [HouseCode\Keycode {\Repeat}{,....}
```

PICBasic Pro includes a number of predefined X-10 function codes. To include these in your PICBasic Pro program, add the following line to the beginning of your program:

```
Include "modedefs.bas"
```

TABLE 10-3   Keycode functions

Function	Operation
unitOn	Turn module on.
unitOff	Turn module off.
unitsOff	Turn all modules off.
lightsOn	Turn all light modules on.
lightsOff	Turn all light modules off.
bright	Brighten light module.
dim	Dim light module.

TABLE 10-4   TW-523 wiring

Wire #	Wire Color	Function
1	Black	Zero-crossing output
2	Red	Common
3	Green	X-10 receive
4	Yellow	X-10 transmit

Table 10-3 shows some keycode functions you can use in your program. The TW-523 has the wiring shown in Table 10-4.

The following is some sample code:

```
house var byte
unit var byte

Include "modedefs.bas"

house = 3      'set housecode D
unit = 7       'set unit to 8

'Turn on unit 7 in house 3
XOUT PORTB.1, PORTB.0,[house\unit, house\unitOn]
```

Speech Synthesizer

This chapter begins your applications. Your first project is a speech synthesizer that may be embedded into another circuit or project to add speech capabilities. With this circuit, you may create a talking robot or an electrical device that can communicate verbally.

Speech synthesizers (or processors) appear in two formats. The first format uses sampled (digitally recorded) speech stored in *read-only memory* (ROM) or *electrically erasable programmable ROM* (EEPROM). The second approach uses phonemes of English to construct words. A phoneme is a speech sound.

Each format has its advantages and disadvantages. Digitally recorded speech has excellent fidelity but has a limited vocabulary due to the large storage capacity required. The phoneme approach has an unlimited vocabulary, but the speech fidelity isn't as good as sampled speech. Even so, the phoneme approach usually suffices as long as a mechanical (robotic type) voice is acceptable. This is the approach you will be using.

The total cost of this project, including the PIC microcontroller, audio amplifier, and speaker should be less than $50.

Although the speech synthesizer chip is capable of producing an unlimited vocabulary, you will not have an unlimited memory in the microcontroller. The finite memory in the microcontroller limits the number of words that can be stored and subsequently spoken, but you may choose any words you want the circuit to speak. You could also interface serial EEPROMs with the microcontroller that can be used to increase word vocabulary.

Speech Chip SPO256

The SPO256 speech synthesizer chip you will be using has been discontinued by General Instruments a number of years ago. However, a good supply of chips is available from Images SI, Inc., as well as a few other distributors (see the suppliers index). The

SPO256 (see Figure 11-1) can generate 59 allophones (the electronic equivalent of English phonemes) plus five pauses (no sound) of various lengths. An allophone table is provided in Table 11-1.

By concatenating (adding) allophones together, we construct words and sentences. This may appear difficult at first, but it is not. Once you get the hang of it, you can turn out complete sentences in a minute or so.

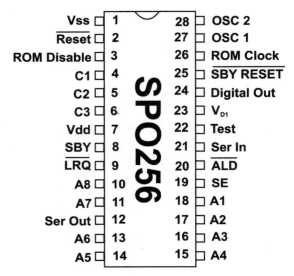

FIGURE 11-1   SPO256 speech generator chip pin output

TABLE 11-1   Allophones

Decimal address	Allophone	Sample word	Duration (milliseconds)
0	PA1	Pause	10
1	PA2	Pause	30
2	PA3	Pause	50
3	PA4	Pause	100
4	PA5	Pause	200
5	OY	Toy	420
6	AY	Buy	260
7	EH	End	70
8	KK3	Cat	120
9	PP	Power	140
10	JH	Judge	140
11	NN1	Pin	140
12	IH	Sit	70
13	TT2	To	140
14	RR1	Pural	170

TABLE 11-1   Allophones *(continued)*

Decimal address	Allophone	Sample word	Duration (milliseconds)
15	AX	Succeed	70
16	MM	My	180
17	TT1	Tart	100
18	DH1	They	290
19	IY	Tee	250
20	EY	Beige	280
21	DD1	Should	70
22	UW1	To	100
23	AO	Aught	100
24	AA	Home	100
25	YY2	Yes	180
26	AE	Pat	120
27	HH1	Him	130
28	BB1	Boy	80
29	TH	They	180
30	UH	Book	100
31	UW2	Food	260
32	AW	Out	370
33	DD2	Don't	160
34	GG3	Pig	140
35	VV	Venom	190
36	GG1	Gotten	80
37	SH	Sharp	160
38	ZH	Azure	190
39	RR2	Train	120
40	FF	Forward	150
41	KK2	Sky	190
42	KK1	Came	160
43	ZZ	Zolu	210
44	NG	Anchor	220
45	LL	Lamb	110
46	WW	Wood	180
47	XR	Pair	360
48	WH	Whine	200
49	YY1	Yes	130
50	CH	Chump	190
51	ER1	Tire	160
52	ER2	Tire	300
53	OW	Beau	240
54	DH2	They	240
55	SS	Best	90
56	NN2	Not	190
57	HH2	Noe	180
58	OR	Pore	330
59	AR	Arm	290
60	YR	Clear	350
61	GG2	Guide	40
62	EL	Paddle	190
63	BB2	Boy	50

A Little on Linguistics

When programming words for the SPO256 speech chip, you must string together the allophones shown in Table 11-1. Words and sentences must end with a pause (silence); if not, the last allophone sent to the chip will drone on continuously.

To pronounce the word "cookie," use the following allophones: KK3, UH, KK1, IY, and PA2. The decimal addresses for the allophones are sent to the SPO256, and this works out to the following numbers: 8, 30, 42, 19, and 1.

The optional data sheet for the SPO256 has an allophone word list containing 200 or so commonly used words (numbers, months, days of the week, and so on). If the word you need isn't on the list, you can make the word up using the allophone list yourself.

The first thing to keep in mind when creating an allophone list for any particular word is that sounds and letters do not have a one-to-one correspondence. You need to spell the words phonetically using the allophone table. For instance, CAT becomes KAT using the allophones KK1 EY TT1 PA1. The decimal addresses for the allophones are 42, 20, 17, and 1. Those are the numbers you plug into your program for it to speak. When the word is programmed in, listen to it as it plays through the SPO256 and, if necessary or desirable, try to improve upon it. In our CAT example, you will find the KK3 allophone makes the word sound better.

Placement of a speech sound within a word can change its pronunciation. For instance, look at the two D's in the word depend. The D's are pronounced differently. The DD2 allophone will sound correctly in the first D position, and the DD1 allophone sounds correctly in the second D position.

General Instruments recommends using a 3.12 MHz crystal at pins 27 and 28. I have used a 3.57 MHz TV color burst crystal on many occasions (due to its availability and the 3.12 MHz unavailability) without any ill effects. The change increases the timbre of the speech slightly.

Interfacing to the SPO256

The pin output and functions of the SPO256 are provided in Table 11-2. The SPO256 has eight address lines (A1 to A8). In your application, you need to access 64 allophones. Therefore, you only need to use address lines A1 to A6. The two other address lines A7 and A8 are tied to ground (0), so any access to the SPO256 address bus will include the address you place on A1 to A6 with lines A7 and A8 equal to 0. Essentially, A7 and A8 add nothing to the address.

Mode Select

Two modes are available for accessing the chip. Mode 0 (SE = 0) will latch an address whenever any of the address pins make a low to high transition. You can think of this as an asynchronous mode.

Mode 1 (SE = 1) latches an address using the ALD pin. When the ALD pin is pulsed low, any address on the lines is latched in. To insure proper synchronization, two pins can tell the microcontroller when the SPO256 is ready for the next allophone address

TABLE 11-2 SPO256 pin functions

Pin #	Name	Function
1	Vss	Ground
2	Reset	Logic 0 resets
	Logic 1 normal operation	
3	ROM Disable	Used with external serial ROM and Logic 1 means it is disabled
4, 5, 6	C1, C2, C3	Output control lines for use with serial ROM
7	Vdd	Power (+5 *volts direct current* [VDC])
8	SBY	Standby (Logic 1 inactive, Logic 0 active)
9	LRQ	Load request (Logic 1 active, Logic 0 inactive)
10, 11, 13, 14,	A8, A7, A6, A5,	Address lines
15, 16, 17, 18	A4, A3, A2, A1	
12	Ser Out	Serial address out for use with serial ROM
19	SE	Strobe enable (Normal set to logic 1 [Mode 1])
20	ALD	Address load (Negative pulse loads address into port)
21	Ser In	Serial in (For use with serial ROM)
22	Test	Grounded for normal operation
23	Vd1	+5 VDC for interface logic
24	Digital Out	Digital speech output
25	SBY Reset	Standby reset (Logic 0 resets)
26	ROM clock	1.56 MHz clock output for use with serial ROM
27	OSC1	XTAL in 3.12 MHz
28	OSC2	XTAL out 3.12 MHz

to be loaded. You will use one of those pins called the SBY pin. The SBY goes high while the chip is enunciating the allophone. As soon as it is completed, the SBY line goes low. This signals the microprocessor to load the next allophone address on lines A1 to A6 and pulse the ALD line low.

The Circuit

The circuit is shown in Figure 11-2. The circuit uses two switches to trigger speech. It is important to realize that the switches provide digital logic signals to the Port A pins, and any circuit that can output Binary 0's and 1's can be used to trigger the circuit to speak. In other words, you don't need to use switches; you may use external logic circuits to place binary signals.

Looking at Figure 11-2, you can see the RA0 and RA1 lines are normally kept high (+5 volts) at binary 1 through the 10K resistor. When a switch is closed, it connects the pin to ground and the line is brought down to (ground) binary 0. I could have arranged the logic signals to the pin(s) to be the opposite. This would have changed a few commands in the program, but functionally it would do the same thing. You choose the logic signal to use based on the circuit you are working with.

The other important point to know is that the five open lines on Port A may be used to trigger up to 31 different speech announcements. The five pins form a five-bit binary number that can be read from Port A. This is hinted at in the program. You will only

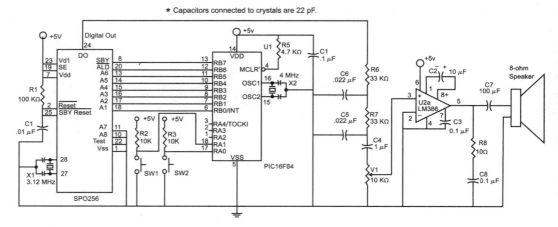

FIGURE 11-2   Speech-generator circuit

TABLE 11-3   Four speech combinations from two numbers

Logic status RA0	RA1	Action
1	1	None—normal state
1	0	Speak word 1
0	1	Speak word 2
0	0	Speak word 3

use two of the five available lines, RA0 and RA1, to jump to three different words. Using a two-bit number, you have four possible combinations, as shown in Table 11-3.

In a similar fashion, a three-bit number allows eight unique combinations, four allows 16, and a five-bit number will allow 31 unique combinations.

In the program, word three is actually a sentence. This simply demonstrates that you are not limited to using single words. The prototyped circuit is shown in Figure 11-3.

The following is a code example:

```
'PICBasic Program 11.1 The SPO256 Talker
SYMBOL TRISB = 134
symbol portb = 6
symbol porta = 5
'Initialize ports
Poke TRISB,128              'Set RB7 as input, Set RB0-RB6 outputs
start:
pause 200                   ' Give a human a chance to press a button(s)
'Check Port A line status could be switch or could be TTL logic signals
peek porta,B0               ' Read Port A
```

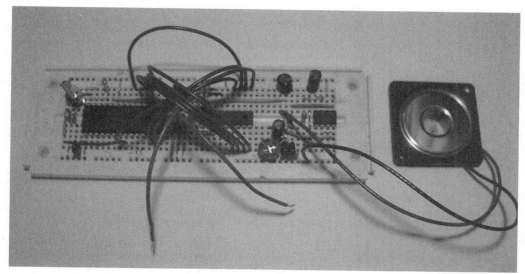

FIGURE 11-3 Prototype circuit

```
If B0 = 0 then three          'Check both lines first (normally b0 = 3)
If bit0 = 0 then hello        'Check line 0 / alternate command if b0 = 2
If bit1 = 0 then world        'Check line 1 / alternate command if b0 = 1
goto start
'
'Say word hello
hello:'It's not just a word, it's a routine
for b3 = 0 to 5                        'loop using number of allophones
lookup b3,(27,7,45,15,53,1),b4         'decimal addresses of allophones
gosub speak                            'speak subroutine
next b3                                'get next allophone
goto start                             'do it again from the beginning
'
'Say word world
world:'Procedure similar to Hello
for b3 = 0 to 4
lookup b3, (46,58,62,21,1),b4
gosub speak
next b3
goto start
'
'Say Sentence 'See you next Tuesday.'
three:'Procedure similar to Hello
for b3 = 0 to 19
lookup b3,(55,55,19,1,49,22,1,11,7,42,55,13,2,13,31,43,2,33,20,1),b4
gosub speak
next b3
goto start
'
speak:'Subroutine to speak allophones
poke portb,b4                   'set up allophone address and bring ALD low
```

```
pause 1                           'pause 1 millisecond for everything to stabilize
high 6                            'Bring ALD high / alternate poke portb, 64
wait:
peek portb,B0                    'Look at port B
if bit7 = 0 then wait            'Check SBY line (0 = talking 1 = finished)
return                           'get next allophone
```

The following program is the PICBasic Pro compiler equivalent:

```
'PICBasic Pro Program 11.2 SPO256 talker
'DEFINE variables
b3 VAR BYTE
b4 VAR BYTE
'Initialize ports
TRISB = 128                           'Set RB7 as input, set RB0 to RB6 as outputs
start:
Pause 200                             'Give a human a chance to press a button(s)
'Check Port A line status could be switches or could be TTL logic signals
IF PORTA = 0 Then three               'Check both lines first (normally = 3)
IF PORTA.0 = 0 Then hello             'Check line 0
IF PORTA.1 = 0 Then world             'Check line 1
GoTo start
'
'Say word hello
hello:'It's not just a word, it's a routine
For b3 = 0 to 5                       'Loop using number of allophones
LookUp b3,[27,7,45,15,53,1],b4        'Decimal addresses of allophones
GoSub speak                           'Speak routine
Next b3                               'Get next allophone
GoTo start                            'Do it again from the beginning
'
'Say word world
world:'Procedure similar to hello
For b3 = 0 to 4
LookUp b3,[46,58,62,21,1],b4
GoSub speak
Next b3
GoTo start
'
'Say sentence                        'See you next Tuesday'
three:'Procedure similar to hello
For b3 = 0 to 19
LookUp b3,[55,55,19,1,49,22,1,11,7,42,55,13,2,13,31,43,2,33,20,1],b4
GoSub speak
Next b3
GoTo start
'
speak:'Subroutine to speak allophones
PORTB = b4                            'Set up allophone address and bring ALD low
Pause 1                               'Pause 1 ms for everything to stabilize
PORTB.6 = 1                           'Bring ALD high
hold:
' Look at port B
IF PORTB.7 = 0 Then hold              'Check SBY line (0 = talking, 1 = finished)
Return                                'Get next allophone
```

Program Differences

Before examining the program functions, let's take a quick look at some of the differences between the PICBasic and PICBasic Pro programs. Primarily, the PICBasic Pro uses square brackets, [ ], in the lookup command, and the PICBasic uses rounded parentheses, ( ). The subroutine "wait" in the PICBasic program was renamed "hold" in the PICBasic Pro program because "wait" is a reserved word in the Pro version used in the serial commands. Other differences in the programs will be illustrated in the following section.

Program Functions

Usually, each program has a little something different than all the other programs you have looked at so far, and these programs are no exception.

Starting from the top, let's look at Port A. In the PICBasic program, you used the Peek command to look at your two switches connected to Port A. The peeked value was stored in the variable B0.

Peek PortA, b0

In the PICBasic Pro program, you can read the port directly, right inside the decision command line as shown here:

```
IF PORTA = 0 Then three     'Check both lines first (normally = 3)
```

By looking at Port A, you can tell if either switch connected to Port A is pressed. Normally, there are two logic highs (+5 volts) on pins RA0 and RA1. Seeing this port as a binary number would look something like XXX00011, where X's mean the pins don't matter and are not available for use (you would typically see these pins as 0). Following the X's are the binary 0's equal to pins RA4, RA3, and RA2. Finally, the binary 1's are equal to RA1 and RA0. The decimal equivalent of this binary number when both these pins are high is three. If you forgot how to read binary numbers, look back to Chapter 7, "PIC 16F84 Microcontroller."

The program interprets the peeked information in a few ways. In the PICBasic program, the peeked value of the port is stored in variable B0.

```
PICBasic Program
If B0 = 0 then three

PICBasic Pro Program
If PortA = 0 then three
```

The only way B0 (PortA) can be zero is if both switches are closed simultaneously. In that case, the program jumps to the routine labeled three. If PortA (B0) does not equal 0, the program continues to the next line.

```
PICBasic Program
If bit0 = 0 then hello

PICBasic Pro program
If PortA.0 = 0 then hello
```

In the PICBasic program line, you are testing the bit 0 value held in the variable B0. This bit-checking ability can only be performed on variables held in the B0 and B1 (or W0) using the PICBasic compiler. Using the PICBasic Pro compiler, you can read the bit status of Port A directly using the command "If PortA.0 = 0 then hello."

The Lookup commands read numbers rather than ASCII codes. To read numbers, leave out the quotation marks after the parentheses. ASCII codes use the quotation marks as in ("H") format.

In the speak subroutine near the end of the program, the PICBasic program uses the Peek command again. This time it looks at the one input line on Port B, the RB7 line. Notice that you can look at the entire Port B (eight bits) even though only one input line exists. The *input/output* (I/O) status of the other lines doesn't impact the usability of the Peek command. When Peek is used on an output line (or port), the result shows us the status of the output line(s) as well as the input lines:

```
peek portb,B0        'Look at port B
if bit7 = 0 then wait    'Check SBY line (0 = talking 1 = finished)
```

After looking at Port B, check the status of the one input line, RB7, using the Bit7 command. In the PICBasic Pro program, you can read the bit off Port B directly using the following command:

```
If PortB.7 = 0 then wait.
```

The input line RB7 is connected to the SBY line of the SPO256. The SBY line stays low while the chip is talking. When it finishes, the line goes high. This tells the PIC microcontroller the SPO256 is ready for the next allophone. While the chip is talking (SBY line low), the program holds in a waiting loop until SBY goes high.

This is the first program in the book that uses Gosub routines. In general, it is recommended not to nest more than three Gosub routines or you stand a good chance of fouling up the stack.

What's a stack? Let's just say it's a pointing register that stores return addresses on top of one another. The stack holds the addresses arranged in a *last in, first out* (LIFO) sequence.

Parts List

The parts list here includes the components outlined in Chapter 1, "Microcontrollers," and Table 11-4 outlines the other necessary items. These components are available from Images SI, Inc., James Electronics, JDR Microdevices, and RadioShack (see the suppliers index).

TABLE 11-4   Additional components

1	SPO256 speech processor
1	8-ohm speaker
1	LM386 audio amplifier
1	10K potentiometer PC mount
3	.1 uF capacitors
2	.022 capacitors
1	1 uF capacitor
1	10 uF capacitor
1	100 uF capacitor
2	Pushbutton switches, normally open
1	100K, 1/4-watt resistor
2	10K, 1/4-watt resistors
2	33K, 1/4-watt resistors
1	10-ohm, 1/4-watt resistor
1	3.12 MHz crystal

Creating a New I/O Port

The speech-generator project from the last chapter demonstrates how a project can quickly gobble up *input/output* (I/O) lines. In complex projects, it's easy to run out of I/O lines, so in this next project you will confront this problem head on.

When one runs out of I/O lines, the first thought is usually to upgrade to a larger PIC microcontroller, such as the 16F873 with 22 I/O lines. Eventually, regardless of the microcontroller chosen, the I/O lines still get used up. Thus, it's to your benefit to learn how to expand existing I/O lines. In this project you will take two or three I/O lines off Port B and expand those to eight output lines. Then using three or four I/O lines off Port B creates eight input lines. Sound good? Read on.

Serial Communication

You will be using serial communication to expand your I/O lines. Serial communication comes in two flavors, synchronous and asynchronous. Synchronous communication uses a clock line to determine when information on the serial line is valid, whereas asynchronous communication doesn't use a clock line. In lieu of a clocking line, asynchronous communication requires start and stop bits in conjunction with strict adherence to timing protocols for its serial communication to be successful. You will use synchronous communication with a clocking line in these projects.

Output First

To create the output lines, you are going to use a serial-to-parallel converter chip, the 74LS164 (see Figure 12-1). This chip reads 8-bit serial data on pins 1 and 2 and outputs the data on eight parallel lines (QA to QH).

If you remember from the command description of the PICBasic and PICBasic Pro language, you have built-in Serial In and Serial Out commands. Unfortunately, you cannot use these Basic commands because their serial format uses stop and start bits, which are necessary in asynchronous (without clock) communication.

FUNCTION TABLE					
Clock	A	B	QA	QB ...	QH
X	X	X	L	L ...	L
L	X	X	Qa	Qb ...	Qh
↑	H	H	H	Qa ...	Qg
↑	L	X	L	Qa ...	Qg
↑	X	L	L	Qa ...	Qg

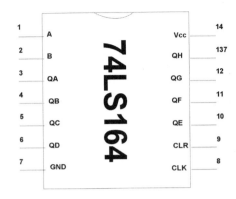

H = High level
L = Low level
X = Irrelevant (any input including transitions)
↑ = Transition from low to high
Qa ... Qg = The level after the most recent ↑ transition of
            the clock; indicates a one-bit shift

FIGURE 12-1   Pinout 74LS164 serial-to-parallel chip

The 74LS164 converter chips use a clock line and do not use or require stop and start bits. Since there is no way to remove these bits from the standard serial Basic commands, you need to program your own serial communication.

Synchronous communication requires a clocking pulse, which determines when the information on the serial line is valid. For the 74LS164, information (with a value of 0 or 1) on the serial line is valid on the low to high transition of the clock pulse.

NOTE: PICBasic Pro users may use the Shiftin and Shiftout commands to achieve similar results. However, these commands do not achieve true synchronous communication because the clock pin has a timing requirement. So, for compatibility, you are creating your own serial communication without the benefits of those commands.

Basic Serial

Serial data is transmitted with the most significant bit (bit 7) first. Since you are writing the serial routine, you could change this and send out the least significant bit (bit 0) first, but stay with this convention for the purposes of the chapter. Figure 12-2 illustrates how the serial data is read by the 74LS164 and parallel information is output.

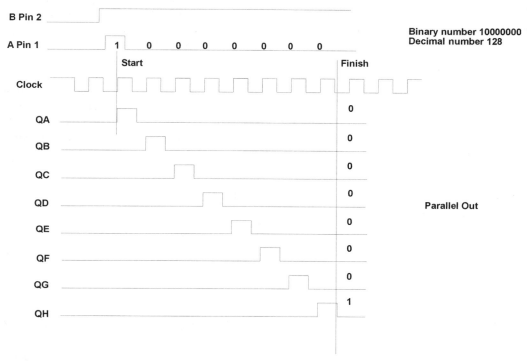

FIGURE 12-2   Serial data in and parallel data out

Line B (pin 2) on the 74LS164 is kept high. This enables you to use line A (pin 1) to send serial data along with the clocking pulse to pin 8. Notice in Figure 12-1, the Function Table, that both lines A and B need to be high for a high bit to be output. You could set either line A or B high and use the other to transmit serial data; it doesn't matter.

Each low to high transition on the clocking line accepts another bit off line A and outputs that bit to QA. All the existing bit information already on the QA to QH lines is shifted one bit to the left. After eight transitions, a new 8-bit number is displayed on lines QA to QH of the 74LS164. In Figure 12-2, binary number 10000000 (decimal number 128) is being transmitted. I chose this number so you can easily see how bit 7, the high bit, shifts down through lines QA to QH.

What isn't immediately evident is that as bit 7 shifts through the QA to QH lines, it brings each QN line high. If a *light-emitting diode* (LED) was attached to each Qn line, you could see bit 7 lighting each LED as it shifted with each transition. Only after eight transitions will bit 7 be in the right position. So, in the first seven transitions, as the serial number is shifting into the 74LS164 parallel register, the number shown on the 8-bit parallel output will be incorrect.

This bit shifting can create chaos in a digital circuit. If the circuit that is connected to the 74LS164 parallel output cannot compensate for this bit shifting, you can correct for this using a second chip, the 74LS373 data octal latch. You will find some circuits can compensate and some others cannot.

Clear Pin

The 74LS164s have an optional pin that can help eliminate some havoc caused by bit shifting. Pin 9 on the 74LS164 is the clear (CLR) pin. It is used to clear whatever binary number exists on the parallel output and brings all lines (QA to QH) low. The CLR pin is active low. For normal operation, this pin is kept high. To clear the number, bring the clear pin low.

First Program

```
'PICBasic Program 12.1
'Slow Serial program for visual testing interface
Symbol TRISB = 134      'Assign Data Direction Register Port B to 134
Symbol PortB = 6        'Assign Variable Port B to decimal value of 6
'Initialize Port(s)
poke TRISB,0            'Set Port B as Output Port
start:
b0 = 128                'Put number 128 (10000000) into b0
gosub serial            'Output the number serially to 74LS164
pause 1000              'wait one second
b0 = 255                'put number 255 (11111111) into b0
gosub serial            'Output the number serially to 74LS164
pause 1000              'wait one second
b0 = 0                  'Put number 0 (00000000) into b0
gosub serial            'Output the number serially to 74LS164
pause 1000              'wait a second
goto start              'do it again

'Serial Out Routine
serial:
pin0 = bit7             'Bring pin0 high or low depending upon bit
pulsout 1, 1            'Bring CLK line high then Low
pause 100               'OPTIONAL Delay ** Remove from program
pin0 = bit6             'Same as above
pulsout 1, 1            'Same as above
pause 100               'OPTIONAL Delay ** Remove from program
pin0 = bit5
pulsout 1, 1
pause 100               'OPTIONAL Delay ** Remove from program
pin0 = bit4
pulsout 1, 1
pause 100               'OPTIONAL Delay ** Remove from program
pin0 = bit3
pulsout 1, 1
pause 100               'OPTIONAL Delay ** Remove from program
pin0 = bit2
pulsout 1, 1
pause 100               'OPTIONAL Delay ** Remove from program
pin0 = bit1
pulsout 1, 1
pause 100               'OPTIONAL Delay ** Remove from program
pin0 = bit0
```

```
pulsout 1, 1
pause 100              'OPTIONAL Delay ** Remove from program
low 1
return
```

For PICBasic Pro, the program is as follows:

```
'PICBasic Pro Program 12.2
'Slow Serial program for visual testing interface
'Initialize variable B0
B0 VAR BYTE
'Initialize port(s)
TRISB = 0             'Set port B as output port
start:
B0 = 128              'Put number 128 (10000000) into b0
GoSub serial          'Output the number serially to 74LS164
Pause 1000            'Wait 1 s
B0 = 255              'Put number 255 (11111111) into b0
GoSub serial          'Output the number serially to 74LS164
Pause 1000            'Wait 1 s
B0 = 0                'Put number 0 (00000000) into b0
GoSub serial          'Output the number serially to 74LS164
Pause 1000            'Wait 1 s
GoTo start            'Do it again
                      'Serial out routine
serial:
PORTB.0 = B0.7        'Bring pin 0 high or low, depending upon bit
PulsOut 1,1           'Bring CLK line high, then low
Pause 100             'Optional delay-remove from program
PORTB.0 = B0.6        'Same as above
PulsOut 1,1           'Same as above
Pause 100             'Optional delay-remove from program
PORTB.0 = B0.5
PulsOut 1,1
Pause 100             'Optional delay-remove from program
PORTB.0 = B0.4
PulsOut 1,1
Pause 100             'Optional delay-remove from program
PORTB.0 = B0.3
PulsOut 1,1
Pause 100             'Optional delay-remove from program
PORTB.0 = B0.2
PulsOut 1,1
Pause 100             'Optional delay-remove from program
PORTB.0 = B0.1
PulsOut 1,1
Pause 100             'Optional delay-remove from program
PORTB.0 = B0.0
PulsOut 1,1
Pause 100             'Optional delay-remove from program
PORTB.1 = 0
Return
```

The schematic shown in Figure 12-3 does not correct for the bit shifting. LEDs are directly connected to the 74164 output lines. The program outputs binary 10000000 (decimal 128), waits a second, outputs 11111111 (decimal 255), waits a second, and then

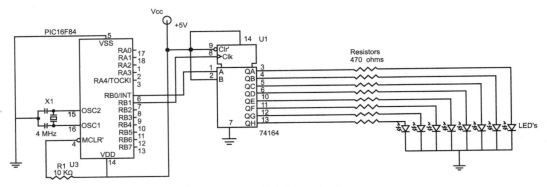

FIGURE 12-3   Schematic using LEDs to show parallel data output

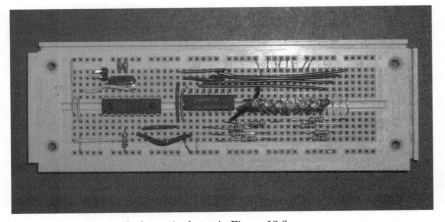

FIGURE 12-4   Prototype circuit of schematic shown in Figure 12-3

outputs 00000000 (decimal 0). Figure 12-4 shows the circuit prototyped on a solderless breadboard.

The first serial program has optional pause commands (Pause 100) after each bit shift that slow the program to enable you to see the bit shifting into and through the register. The Pause commands ought to be removed when using the serial routine in a real application. If you remove the Pause command(s) from the program(s), as shown in the following two programs, recompile the program and upload it into the 16F84. When this program is run, the bit numbers shift so quickly, you will not see the bit shifting occurring, but an electronic circuit connected to the outputs of the 74LS164 will.

This is why I wrote the slow serial programs, so the bit shifting becomes clearly evident. The following two programs are the same as the previous programs with the exception of having the "pause 100" removed after each bit shift and having pin RB2 activate the 74LS373 data octal latch.

Bit Shift Correcting

If you are interfacing with a digital circuit that would be sensitive to the bit shifting, you can block the bit shifting by using a second chip: the 74LS373 data octal latch (see Figure 12-5). The octal latch is placed between the output of the 74LS164 and the LEDs (see Figure 12-6). Remember in your testing circuit that the LEDs represent your digital circuit. Tie the OC (Output Enable) pin to ground and use the C (Enable Latch) pin

The eight latches of the LS373–when C is high, the Q outputs follow the D inputs. When C is low, the output will be latched at the current data levels.

1	$\overline{QC}$		Vcc	20
2	1Q		8Q	19
3	1D		8D	18
4	2D		7D	17
5	2Q		7Q	16
6	3Q		6Q	15
7	3D		6D	14
8	4D		5D	13
9	4Q		5Q	12
10	GND		C	11

74LS373

FUNCTION TABLE			
Output Enable	Enable Latch	D	Output
L	H	H	H
L	H	L	L
L	L	X	Qo
H	X	X	Z

C = Enable Latch

$\overline{OC}$ = Output Enable

FIGURE 12-5   Octal data latch used to nullify bit shifting in output

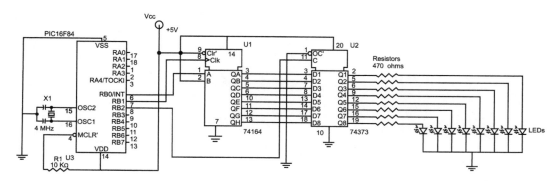

FIGURE 12-6   Schematic using LEDs to show parallel data output with 74LS373 data octal latch to nullify bit shifting in output

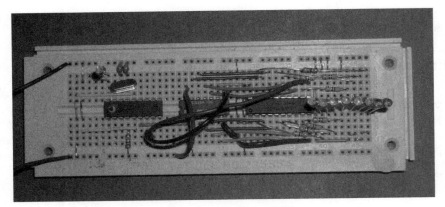

FIGURE 12-7   Prototype circuit of schematic shown in Figure 12-6

(Pin 11 on 74LS373) to control the data through the 74LS373. Figure 12-7 shows this circuit prototyped on a solderless breadboard.

Data is placed on the D inputs (1D, 2D . . . 8D). When you want the data to appear on the Q outputs (1Q, 2Q . . . 8Q), raise the C pin momentarily.

The program inputs the data serially to the 74LS164, and the data appears on the parallel out lines, with bit shifting occurring on the inputs of the 74LS373. When the bit shifting is finished and the binary number has settled (eight shifts), raise the Enable Latch (pin 11) of the 74LS373 that lets the parallel data flow from the input pins to the output pins. Then Enable Latch (pin 11) is lowered, leaving the parallel data latched in.

As the bits shift, the bit shifting is blocked from the LEDs. Only when the entire byte has been transmitted should you raise pin 11 (C pin) on the 74LS373, letting the byte information through to your LEDs and be latched in.

```
' PICBasic Program 12.3 Serial program with latch

Symbol TRISB = 134     'Assign Data Direction Register Port B to 134
Symbol PortB = 6       'Assign Variable Port B to decimal value of 6


'Initialize Port(s)
poke TRISB,0           'Set Port B as Output Port
start:
b0 = 128               'Put number 128 (10000000) into b0
gosub serial           'Output the number serially to 74164
pause 1000             'wait one second
b0 = 255               'put number 255 (11111111) into b0
gosub serial           'Output the number serially to 74164
pause 1000             'wait one second
b0 = 0                 'Put number 0 (00000000) into b0
gosub serial           'Output the number serially to 74164
pause 1000             'wait a second
```

```
goto start                'do it again
'Serial Out Routine
serial:
pin0 = bit7               'Bring pin 0 high or low depending upon bit
pulsout 1, 1              'Bring CLK line high then Low
pin0 = bit6               'Same as above
pulsout 1, 1              'Same as above
pin0 = bit5
pulsout 1, 1
pin0 = bit4
pulsout 1, 1
pin0 = bit3
pulsout 1, 1
pin0 = bit2
pulsout 1, 1
pin0 = bit1
pulsout 1, 1
pin0 = bit0
pulsout 1, 1
low 1
high 2                    'Raise enable latch on 74LS373
low 2                     'Lower enable latch on 74LS373
return
```

The following program is the PICBasic Pro version of the serial output program without using the Pause command between bit shifts:

```
'PICBasic Pro Program 12.4 Serial program with latch
'Initialize variable B0
B0 VAR BYTE
'Initialize port(s)
TRISB = 0                 'Set Port B as output port
start:
B0 = 128                  'Put number 128 (10000000) into b0
GoSub serial              'Output the number serially to 74LS164
Pause 1000                'Wait 1 s
B0 = 255                  'Put number 255 (11111111) into b0
GoSub serial              'Output the number serially to 74LS164
Pause 1000                'Wait 1 s
B0 = 0                    'Put number 0 (00000000) into b0
GoSub serial              'Output the number serially to 74LS164
Pause 1000                'Wait 1 s
GoTo start                'Do it again
'Serial out routine
serial:
PORTB.0 = B0.7            'Bring pin 0 high or low, depending upon bit
PulsOut 1,1               'Bring CLK line high, then low
PORTB.0 = B0.6            'Same as above
PulsOut 1,1               'Same as above
PORTB.0 = B0.5
PulsOut 1,1
PORTB.0 = B0.4
PulsOut 1,1
PORTB.0 = B0.3
PulsOut 1,1
PORTB.0 = B0.2
PulsOut 1,1
```

```
PORTB.0 = B0.1
PulsOut 1,1
PORTB.0 = B0.0
PulsOut 1,1
PORTB.1 = 0
PORTB.2 = 1          'Raise enable latch on 74LS373
PORTB.2 =0           'Lower enable latch on 74LS373
Return
```

Input I/O

You shall continue by expanding four I/O lines off the PIC microcontroller to function like eight input lines. To accomplish this, you will use a 74LS165 parallel-to-serial converter chip (see Figure 12-8). The parallel-in, serial-out diagram is shown in Figure 12-9. Eight-bit parallel information is placed on the 74LS165 eight (A to H) input lines. The shift load line is brought low momentarily to load the parallel information into the chip's registers. The clock inhibit line is then brought from high to low, allowing the information to be serially output from pin QH, in synchronization with the clocking pulses provided on pin 2 (CLK). Note an inverted serial signal is also available from pin 7.

To functionally test the chips, circuit, and program, you will input an 8-bit number to the 74LS165 using switches and resistors (see Figure 12-10). This binary number created with the switches and resistors will be serially shifted out of the 74LS165 into variable B0. The number in B0 will then be serially sent out to the 74LS164 for display. Figure 12-11 shows this circuit prototyped on two solderless breadboards.

You could reduce one of the I/O lines taken up by the 74LS165 if you shared the clocking line of the 74LS164 with the 74LS165. I want to keep the programming as

FUNCTION TABLE				
Shift/ Load	Clock	Clock Inhibit	Parallel A .... H	Output Qh
L	X	X	a...h	h
H	L	L	X	Qho
H	↑	L	X	Qgn
H	↑	L	X	Qgn
H	X	H	X	Qho

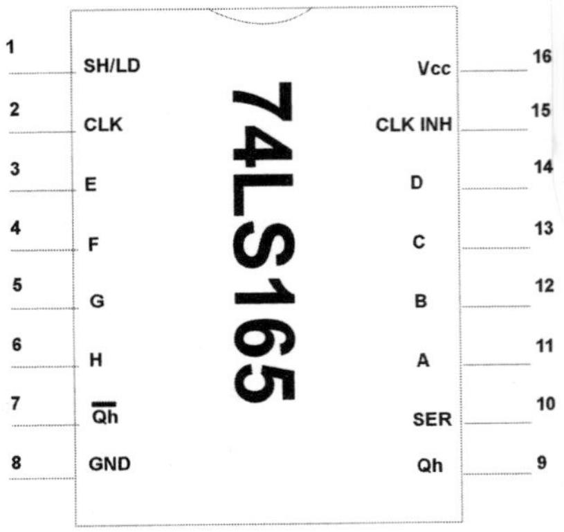

	74LS165	
1	SH/LD	Vcc	16
2	CLK	CLK INH	15
3	E	D	14
4	F	C	13
5	G	B	12
6	H	A	11
7	Q̄h	SER	10
8	GND	Qh	9

H = High level
L = Low level
X = Irrelevant (any input including transitions)
↑ = Transition from low to high

FIGURE 12-8   Pin out for the 74LS165 parallel-in, serial-out chip

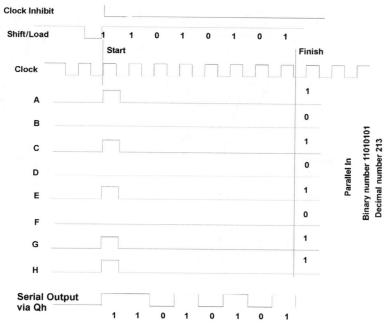

FIGURE 12-9   Parallel data in and serial data out

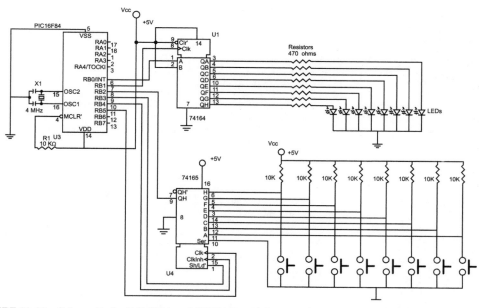

FIGURE 12-10   Schematic for 74LS164 and 74LS165 serial transmission and receiving

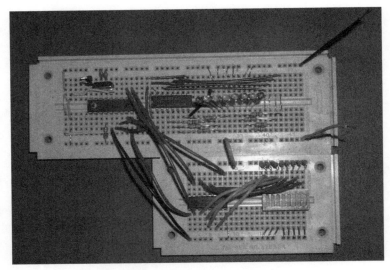

FIGURE 12-11 Prototype circuit of schematic shown in Figure 12-10

straightforward as possible so I won't do this, but you should be aware that a clock-sharing feature could be implemented, and this would reduce the required lines to three.

In addition, the Serial Out and Serial In routines could be layered and merged together, but in order to keep the program as straightforward as possible the Serial Out and Serial In routines are kept separated.

```
' Serial I/O Interface PICBasic Program 12.5
Symbol TRISB = 134     'Assign Data Direction Register Port B to 134
Symbol PortB = 6       'Assign Variable Port B to decimal value of 6
'Initialize Port(s)
poke TRISB,4           'Set Port B Pin2 = input/rest output
Low 5                  'Set CLK low
High 4                 'Bring CLK Inhibit high
High 3                 'Bring Shift/Load High
start:

gosub serial_in        'Get number from 74165
gosub serial_out       'Send it out on 74164
pause 1000             'Wait so I can see it
goto start             'do it again

'Serial In Routine
serial_in:
pulsout 3,1            'Bring Shift/Load down momentarily
low 4                  'Bring CLK inhibit low
bit7 = pin2            'Load bit into B0
pulsout 5,1            'Bring clk pin high then low
bit6 = pin2            'Same as above
pulsout 5,1
```

```
bit5 = pin2
pulsout 5,1
bit4 = pin2
pulsout 5,1
bit3 = pin2
pulsout 5,1
bit2 = pin2
pulsout 5,1
bit1 = pin2
pulsout 5,1
bit0 = pin2
high 4                 'Bring CLK Inhibit High
return

'Serial Out Routine
serial_out:
pin0 = bit7            'Bring Pin 0 high or low depending upon bit
pulsout 1, 1           'Bring CLK line low then high
pin0 = bit6            'Same as above
pulsout 1, 1           'Same as above
pin0 = bit5
pulsout 1, 1
pin0 = bit4
pulsout 1, 1
pin0 = bit3
pulsout 1, 1
pin0 = bit2
pulsout 1, 1
pin0 = bit1
pulsout 1, 1
pin0 = bit0
pulsout 1, 1
low 1
return
```

The following is the equivalent PICBasic Pro program:

```
' Serial I/O Interface PICBasic Pro Program 12.6
BO VAR BYTE
'Initialize Ports(s)
TRISB = 4              'Set port B Pin 2 = input/rest output
Low PORTB.5            'Set CLK low
High PORTB.4           'Bring CLK inhibit high
High PORTB.3           'Bring shift/load high
start:
GoSub serial_in        'Get number from 74LS165
GoSub serial_out       'Send it out on 74LS164
Pause 1000             'Wait so I can see it
GoTo start             'Do it again
'Serial In Routine
serial_in:
PulsOut PORTB.3,1      'Bring shift/load down momentarily
Low PORTB.4            'Bring CLK inhibit low
BO.7 = PORTB.2         'Load bit into B0
PulsOut PORTB.5,1      'Bring CLK pin high, then low'
BO.6 = PORTB.2         'Same as above
PulsOut PORTB.5,1
BO.5 = PORTB.2
```

```
PulsOut PORTB.5,1
BO.4 = PORTB.2
PulsOut PORTB.5,1
BO.3 = PORTB.2
PulsOut PORTB.5,1
BO.2 = PORTB.2
PulsOut PORTB.5,1
BO.1 = PORTB.2
PulsOut PORTB.5,1
BO.0 = PORTB.2
High PORTB.4             'Bring CLK inhibit high
Return
'Serial Out Routine
serial_out:
PORTB.0 = BO.7          'pin 0 high or low depending upon bit
PulsOut PORTB.1, 1      'Bring CLK line low, then high
PORTB.0 = BO.6          'Same as above
PulsOut PORTB.1, 1      'Same as above
PORTB.0 = BO.5
PulsOut PORTB.1, 1
PORTB.0 = BO.4
PulsOut PORTB.1, 1
PORTB.0 = BO.3
PulsOut PORTB.1, 1
PORTB.0 = BO.2
PulsOut PORTB.1, 1
PORTB.0 = BO.1
PulsOut PORTB.1, 1
PORTB.0 = BO.0
PulsOut PORTB.1, 1
PORTB.1 = 0
Return
```

The schematic shows that the parallel input lines to the 74LS165 are connected to Vcc through 10,000-ohm resistors. This puts a binary 1 on each input line. The switches used in this project are an eight-position dip switch. Each switch connects the bottom of each resistor to ground. When the switch is closed, it brings that particular pin to ground, or binary 0. By opening and closing the switches, you can write a binary number on the parallel input. This information is automatically retrieved from the 74LS165 serially by the PIC microcontroller.

The PIC microcontroller takes the number retrieved from the 74LS165 and displays it on the 74LS164. Although this project may appear trivial, it is not. Using this information, you can set up serial communication routines with other chips and systems. In upcoming chapters, you will use a serial routine to communicate to a serial *analog-to-digital* (A/D) converter and use the Basic Serout command to generate *liquid crystal display* (LCDs).

Compatibility Issues

PICBasic and PICBasic Pro programs are more compatible than what previous programs have illustrated. For instance, in the last PICBasic program, commands like "high 4," "low 4," and "pulsout 5,1" were changed to "high portb.4," "low portb.4," and

TABLE 12-1   Port mapping

Number of microcontroller pins	0 to 7	8 to 15
8-pin	GPIO	GPIO
18-pin	Port B	Port A*
28-pin (except 14C000)	Port B	Port C
28-pin 14C000	Port C	Port D
40-pin	Port B	Port C

*If a port does not have eight pins, such as Port A, only the pin numbers that exist may be used.

TABLE 12-2   Additional components

74LS164	Serial-to-parallel *integrated circuit* (IC)
74LS165	Parallel-to-serial IC
8	10K, 1/4-watt resistors
1	Eight-position dip switch

"pulsout portb.5,1" in the PICBasic Pro program. In the Pro version, you identify pins by their port name and number, allowing you to use the command on any available port on compatible microcontrollers. However, this is not always a necessity.

Using the Pro compiler, if a pin is a constant between 0 and 15 (or a variable that contains a number between 0 and 15), it is physically mapped to the microcontroller's I/O ports. For your 18-pin 16F84 microcontroller, numbers 0 to 7 are mapped to Port B pins 0 to 7, just like the PICBasic compiler. Pins 8 to 15 are mapped to Port A pins 0 to 7. Since only five lines are available on Port A, only numbers 8 to 12, which correspond to Port A pins 0 to 4, will have any effect. Table 12-1 shows how the ports are mapped depending upon how many pins the microcontroller has.

In future programs, the programs will be written using these compatibility guidelines. In some cases, the only difference between the PICBasic and PICBasic Pro version will be defining the variables used.

Parts List

The parts here are the same as Chapter 1, "Microcontrollers," and Chapter 8, "Reading I/O Lines," for a programming challenge. Table 12-2 lists the other necessary components. They are available from the Images Company, James Electronics, JDR Microdevices, and RadioShack (see the suppliers index).

Liquid Crystal Display (LCD)

One thing your microcontrollers lack is a display that could provide you with information. A simple alphanumeric display to output text messages and numeric values of variables could enhance the versatility and usefulness of your microcontrollers.

One solution would be a standard video display, but bitmapping graphics and text would demand too much computational overhead. A better solution is an alphanumeric *liquid crystal display* (LCD). Many different types of LCD displays are available on the market. Some higher-end LCD displays support bitmapped graphics for those who may have a need for them, whereas others, like the LCDs you will use, just display letters and numbers (alphanumeric characters) through a simple serial interface.

Your LCD provides two lines, with 40 characters on each line. However, only the first 16 characters on each line are visible in the display. The remainder of the characters are offscreen.

The LCD displays the alphanumeric information it receives via a standard serial format (RS-232). The data rate may be set to 300, 1,200, 2,400, or 9,600 baud (true or inverted). To make matters even better, you only need to employ a single command to output alphanumeric data to the serial LCD.

Serout Command's RS-232 Serial Communication

The compiler's Serout command performs serial, asynchronous communication. Asynchronous means the data flow is time dependent. This communication protocol requires strict time control and the framing of each bit transmitted. The reason for this is because no clock line tells the microcontroller when the information on the serial line is valid. Thus, the time elapse, beginning at the start bit, is the synch factor.

In the previous serial examples in Chapter 12, "Creating a New I/O Port," you had a clock line. As stated earlier, serial communication with a clock line is called synchronous communication. The time between bits being transmitted or received can vary widely, from microseconds to days. The bit becomes valid only when the clock line clocks in the bit.

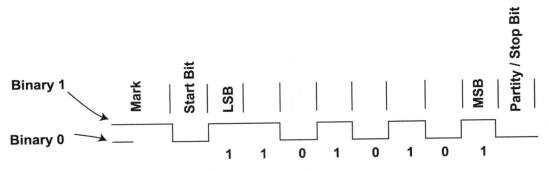

FIGURE 13-1   Standard serial output with start and stop bits

Communicating asynchronously (not in sync or without a clock) in addition to the strict time frame requires the use of start and stop bits. As the name implies, the start bit informs the receiver that a byte of information is about to be transmitted. The start bit is followed by the bit information (usually 8 bits but sometimes 7 bits), which is followed by the stop bit.

Figure 13-1 illustrates a typical serial data communication. Although the line is idle, it is said to be in a mark or marking condition. The mark is a binary 1 that may be a positive voltage or current. Binary 0 is sometimes referred to as a space; it may be a zero voltage or current condition.

To initiate communication, the transmitter sends out a start bit (which brings the line low). Next, the eight data bits are sent. Notice that in this serial communication the *least significant bit* (LSB) is sent first and the *most significant bit* (MSB) is sent last. Note that this is the opposite of the previous synchronous serial communication examples.

Using a standard 8-bit data package plus one stop bit and one start bit equals a total of 10 bits. At 2,400 baud (bits per second), 240 bytes are transmitted each second. At 9600 baud, a maximum of 960 bytes can be transmitted per second.

The frequency (baud rate) of asynchronous communication must be strictly adhered to. Because there isn't a clock line, the next bit in the sequence must be on the line at precise increments of time that are determined by the baud rate. For instance, at 9600 baud, each bit is on the line for 104 microseconds (μsec) and must be captured by the receiver in that time.

In order for things to work correctly, the transmitter and receiver baud rate cannot vary by more than 5 percent of the ideal baud rate. If their respective rates vary by more than that, during the course of 10 bits being transmitted and received they might (not always) fall out of sync by more than a bit (104 μsec). This sync error can corrupt the entire byte received.

Many people looking at a variance of 5 percent don't see how such a small variance could cause a problem. Let's do the math really quickly and see how it happens.

Let's assume the transmitter is 5 percent faster than the ideal rate, and the receiver is 5 percent slower than the ideal rate. For your example, the rate is 2,400 baud, which works out to a time of 416.6 μsec per bit. The transmitter is transmitting at 2,520 baud (2,400 plus 5 percent), or 396.8 μsec per bit. The receiver is receiving at 2,280 baud (2,400 minus 5 percent), or 438.5 μsec per bit. The timing difference is 41.7 μsec per bit. Multiply this difference by the 9 bits and the timing falls out of sync by 375.3 μsec. This is just a touch shorter than the standard bit length of 416.6 μsec at 2,400 baud, so in this case the integrity of the transmission remains valid. But it becomes obvious that if the baud rate varies by any more than 5 percent, the serial communication will fall out of sync by the tenth bit and be corrupted.

Error-Detection Algorithms

Full-featured communication packages may contain algorithms that help prevent data corruption if the asynchronous time frame varies. Error-detection algorithms have not yet become part of the PICBasic or the PICBasic Pro's Serout RS-232 communication routines.

Parity

The stop bit is used to check each byte transmitted using a process known as parity, which may be odd, even, or none. The serial transmission consists of a sequence of binary 1's and 0's. If you choose even parity, the receiver would count how many binary 1's are transmitted. If the number of 1's transmitted is an even number, the stop bit would be made a binary 0, keeping the number of binary 1's even. On the other hand, if the number of binary 1's transmitted is equal to an odd number, the stop bit would be made a binary 1 to make it an even number.

If parity is set to even and an odd number of binary 1's is received, this is known as a parity error and the whole byte of information is thrown out. Parity errors may be caused by power surges, bad communication lines, and poor interface connections. The problems become more pronounced at the faster baud rates.

Serial Format: Mode and Baud Rate

The PICBasic and PICBasic Pro compilers have a few standard formats and speeds available, with baud rates at 300, 1,200, 2,400, and 9,600. Data is sent as eight data bits, no parity, and one stop bit. The mode can be true or inverted. See Chapter 9, "PICBasic Language Reference," for more details on Serin and Serout commands.

XTAL Clock, Please

When I first started using the Serin and Serout commands, they wouldn't work properly. After much hair pulling, I discovered I had used a 3.57 MHz crystal instead of a 4.0 MHz crystal on the PIC 16F84. As soon as I changed the crystal, the Serin and

Serout commands worked perfectly. Later I tried a 4.0 MHz ceramic resonator for the oscillator that also worked properly.

4.0 MHz Clock Limitations

The PICBasic and PICBasic Pro compilers can easily program microcontrollers at 9,600 baud. However, when the microcontroller is operating at a standard clock speed of 4.0 MHz, it is hovering on the limit of acceptable 9,600-baud serial communication. To put it more plainly, 9,600 baud serial communication could easily fail when using a 4.0 MHz Xtal. When using a standard 4.0 MHz clock, I would limit serial communication to a maximum speed of 2,400 baud. If you want to communicate at 9,600 baud, here are a few solutions.

If you are programming using the PICBasic Pro compiler, you can purchase a faster 16F84 microcontroller chip (20 MHz) and a corresponding faster Xtal for the clock. Then define the new clock speed you are using at the beginning of the PICBasic Pro program.

You can use the following Xtal speeds with the DEFINE OSC command: 8, 12, 16, and 20 MHz. Any of these higher clock speeds will insure successful 9,600 baud communication. To inform the compiler that you are using a faster-than-4.0-MHz clock, enter the following line at the beginning of your program:

```
DEFINE OSC X      'Where X is your Xtal speed in MHz
```

So if you are using a 16 MHz Xtal, your command line would read "DEFINE OSC 16." By informing the compiler that you are jumping up to a faster clock speed, the compiler will automatically adjust the timing of the serial communications in the program for you, and you will not have any problem communicating at 9,600 baud.

Keep in mind that you don't have to exactly match the microcontroller speed to the Xtal. For instance, a 20 MHz 16F84 microcontroller will work with 4, 8, 10, 16, and 20 MHz Xtal crystals.

A solution for the PICBasic compiler users is also possible, but a little more convoluted. What you can do is use a 20 MHz microcontroller with a 16 MHz crystal. Using this microcontroller/clock combination will make every command execute at four times its normal rate. Then write your serial communication code for 2,400 baud; it will actually execute at 9,600 baud (4 × 2,400 = 9,600). Keep in mind when doing this that all commands operate faster, so a one-second-delay Pause 1000 will actually execute in $\frac{1}{4}$ of a second. Adjust your commands accordingly.

Three-Wire Connection

The LCD module requires only three wires to function: a +5-volt red wire, a ground (GND) black wire, and a yellow or white serial input line (see Figure 13-2). The communication baud rate of the LCD module may be set to 300, 1,200, 2,400, or 9,600 baud using jumpers J1, J2, and J3 on the back of the LCD (see Figure 13-2). The serial format is eight data bits, one stop bit, and no parity. The baud rate (and mode) set for the LCD must match the baud rate (and mode) used in your program.

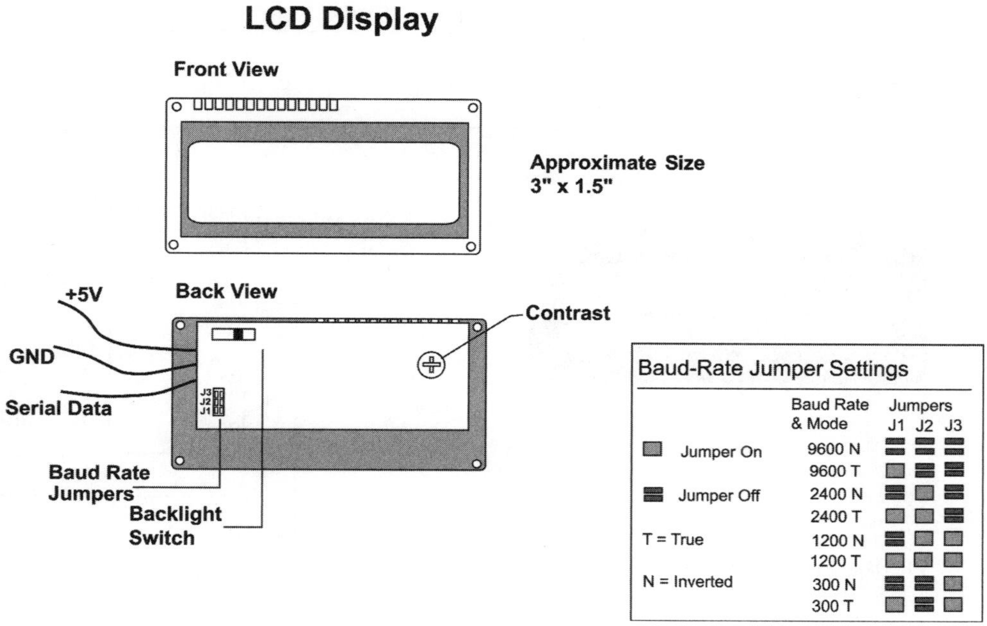

LCD Display

Front View

**Approximate Size
3" x 1.5"**

Back View

+5V

GND

Serial Data

Baud Rate
Jumpers

Backlight
Switch

Contrast

Baud-Rate Jumper Settings

		Baud Rate & Mode	Jumpers J1	J2	J3
▦	Jumper On	9600 N			
		9600 T			
▬	Jumper Off	2400 N			
		2400 T			
T = True		1200 N			
		1200 T			
N = Inverted		300 N			
		300 T			

FIGURE 13-2   Front and back views of the LCD display

Our first program prints out the message "Hello World" on the LCD. The cursor (printing position) automatically moves from left to right. The program is set to a baud rate of 2,400, with the mode inverted. Again, the LCD must be set to the same 2,400 baud rate and inverted mode to receive the incoming serial data.

Set the baud rate and mode on the LCD to 2,400 baud inverted using jumpers J1 through J3 on the back of the LCD. The schematic is shown in Figure 13-3.

```
' PICBasic LCD Test Program 13.1
pause 1000                          'Wait 1 second for LCD
serout 1, N2400, (254, 1)           'Clear screen
pause 1                             'Wait 1 millisecond
serout 1, N2400, ("Hello World!")   'Print message
end
```

The following program is the PICBasic Pro equivalent:

```
' PICBasic Pro LCD Test Program 9.2
pause 1000                          'Wait 1 second for LCD
serout 1, N2400, [254, 1]           'Clear screen
pause 1                             'Wait 1 millisecond
serout 1, N2400, ["Hello World!"]   'Print message
end
```

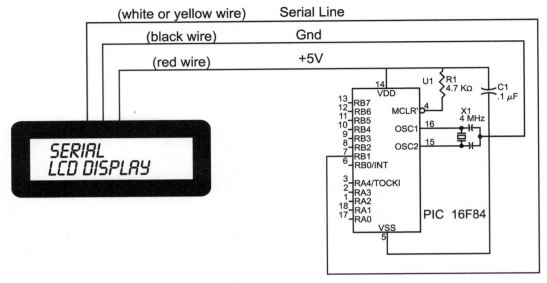

FIGURE 13-3 Schematic of LCD serial display to PIC 16F84 microcontroller

I kept these programs small to show how easy it is to print a message on the LCD from the PIC microcontroller. The difference between the PICBasic and PICBasic Pro programs is the brackets used in the command lines. The PICBasic compiler uses parentheses, ( ), and the PICBasic Pro compilers uses brackets, [ ].

In the program, line 2 "serout 1, N2400, (254,1)" is an LCD command. The LCD screen has a variety of commands, and the most popular 11 commands are listed in Table 13-1. All commands must be prefixed with decimal number 254. The display will treat any number following the 254 prefix as an instruction.

TABLE 13-1 Instruction codes for LCDs

Code	Instruction
1	Clear screen.
2	Send cursor to top-left position (home).
8	Blank without clearing.
12	Make cursor invisible/restore display if blanked.
13	Turn on visible blinking cursor.
14	Turn on visible underline cursor.
16	Mover cursor one character left.
20	Move cursor one character right.
24	Scroll display one character left (all lines).
28	Scroll display one character right (all lines).
192	Move cursor to first position on second line.

Positioning the Cursor

The cursor may be positioned anywhere on the LCD screen by using the following command: 254, position number. The position number can be determined by looking at Figure 13-4. If you wanted to move the cursor to the second line position for 10, you would use the following command:

```
serout 1, N2400, (254, 201)
```

Off-Screen Memory

As stated previously, each line on the LCD holds 40 characters, and only the first 16 characters are displayed on the LCD screen. You can use the scroll commands to view the nondisplayed text on each line.

This second program illustrates how to move the cursor to the second line to print a second line of text and then scroll the entire display message back and forth (see Figure 13-5).

```
'PICBasic LCD Test Program 13.3
Pause 1000
start:
Serout 1, N2400, (254, 1)
Pause 2
Serout 1, N2400, ("Wherever you go.")
Serout 1, N2400, (254, 192)
Pause 2
Serout 1, N2400, ("There you are.")
'Scroll Right
For B0 = 0 to 16
Serout 1, N2400, (254, 28)
Pause 200
Next B0
'Scroll Left
```

LCD Display Screen

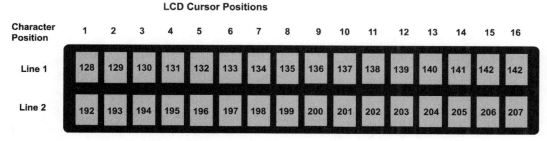

FIGURE 13-4   LCD screen and cursor positions

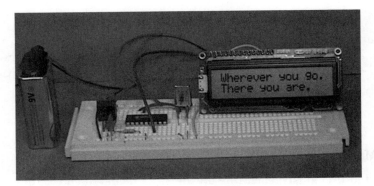

FIGURE 13-5   LCD message

```
For B0 = 0 to 16
Serout 1, N2400, (254, 24)
Pause 150
Next B0
Goto start
End
```

Here is the equivalent PICBasic Pro Program:

```
'PICBasic Pro LCD Test Program 13.4
B0 VAR BYTE
Pause 1000
start:
Serout 1, 4, [254, 1]      'Clear Screen
Pause 15
Serout 1, 4, ["Wherever you go."]
Serout 1, 4, [254,192]
Pause 2
Serout 1, 4, ["There you are."]
'Scroll Right
For B0 = 0 to 16
Serout 1, 4, [254, 28]
Pause 200
Next B0
'Scroll Left
For B0 = 0 to 16
Serout 1, 4, [254, 24]
Pause 150
Next B0
Goto start
End
```

PICBasic Pro Project: LCD Module

This particular LCD project is only suitable for PICBasic Pro compiler users. The LCD modules used so far in this chapter have a built-in serial interface. Although this simplifies using the LCD, it is more expensive due to the serial interface. You can purchase inexpensive LCD modules for $9.95 and connect them directly to your microcontroller. This is a considerable cost savings in comparison to the $38.95 for a serial LCD.

The LCD module for this project must have a Hitachi 44780 controller or equivalent. This controller appears pretty standard since all the LCD modules I've checked so far work.

These LCDs usually have either a 14-pin (no backlight) or 16-pin, (with backlight) single-line or dual-line header on one edge. You need to have the data sheet for the LCD module to find the correct pins to connect the microcontroller. The pins you need to find on the LCD module are as follows:

- **Vss** Ground
- **Vdd** +5 volts
- **VO** Input voltage for LCD
- **RS** High or low (high: data, low: instruction code)
- **R/W** High/low (high: read mode, low: write mode)
- **E** Chip enable
- **DB0–DB7** Data bits 0 through 7

Figure 13-6 details connecting an LCD to the 16F84 microcontroller using the LCD-04 Backlight module from Images SI for $9.95. The microcontroller interfaces to the LCD module using a 4-bit data bus. On this particular LCD module, a 16-pin single-line

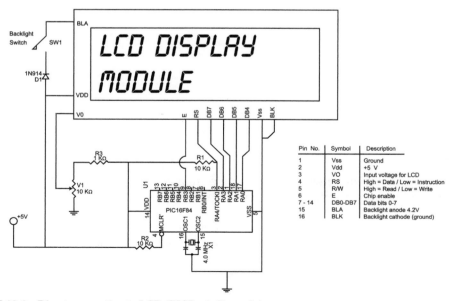

FIGURE 13-6 Direct connection to LCD (PICBasic Pro only)

header is located at the top edge of the LCD module. Pin hole number 1 is at the upper left-hand corner of the module and increments by one to the right. I soldered pin headers (.100 × .100 center to center) into these holes.

Using the LCD Module for Display

To display messages on the LCD, use the LCDOUT command. Here's an example:

```
'PICBasic Pro Program 13.5
Pause 1000          'Pause to allow LCD to set up
LCDOut 254,1        'Clears display
LCDOut "Hello"      'Prints hello
End
```

This program illustrates how easy it is to print a message on your LCD screen. Notice that LCD instruction commands are preceded with the number 254 as you had done with the serial LCD modules. In addition, the LCD instruction set (commands) for the serial LCD modules is the same.

This next program prints and scrolls the message:

```
'PICBasic Pro LCD Test Program 13.6
B0 VAR BYTE
Pause 1000
start:
LCDOUT 254, 1       'Clear Screen
LCDOUT "Wherever you go."
LCDOUT 254, 192
LCDOUT "There you are."
'Scroll Right
For B0 = 0 to 16
LCDOUT 254, 28
Pause 200
Next B0
'Scroll Left
For B0 = 0 to 16
LCDOUT 254, 24
Pause 200
Next B0
Goto start
End
```

Parts List

The parts needed here are the same components as Chapter 1, "Microcontrollers," and the additional commands are as follows:

- LCD-02 serial LCD module (with Backlight), $38.95
- LCD-04 LCD module (with backlight) (Pro compiler only), $9.95

They are available from the Images Company, James Electronics, and JDR Microdevices (see the suppliers index).

Reading Resistive Sensors

The Pot command is powerful in scope and capabilities. It enables users to easily and accurately read resistive components and sensors. The command can read resistive values up to approximately 50,000 ohms (50 kilo-ohms) in a single program line. This command was first reviewed in Chapter 9, "PICBasic Language Reference." The Pot command's structure is as follows:

```
Pot Pin, Scale, Var
```

This command reads a potentiometer or another resistive component on the *Pin* specified (see Figure 14-1). For PICBasic compiler users, you may choose any of the Port B pins, 0 to 7, to use with this command. PICBasic Pro compilers may use these pins plus additional pins on other *input / output* (I/O) ports.

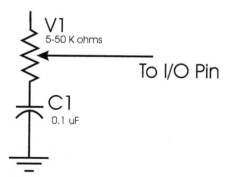

FIGURE 14-1 Resistive sensor (potentiometer) and capacitor connected to microcontroller I/O pin

Resistance is measured by timing the discharge of a capacitor through the resistor, usually 5K to 50K. Scale is used to adjust the varying *resistor-capacitor* (RC) constants. For large RC constants, set Scale to 1. For small RC constants, set Scale to its maximum value of 255. Ideally, if Scale is set correctly, the variable Var will be set to 0 at minimum resistance and to 255 at maximum resistance.

R/C Values

Figure 14-2 graphs the decimal value generated by resistance for four common capacitor values. This chart should be used as a guide to determine the capacitance one could use for the resistance range one needs to read. For instance, using a .1 µF capacitor, the decimal values from 50 to 250 are equal to resistance values of 700 ohms to 3,500 ohms. For a .022 µF capacitor, the same decimal values equal a resistance range of 3,500 ohms to 21 kilo-ohms. Once a range is chosen, the scale factor in the command line may be used to fine-tune the response.

Scale

Scale is determined experimentally. Set the device or transducer to measure at a maximum resistance and read it with a scale set to 255. Under these conditions, the numeric value of *Var* produced will be an approximate ideal value for Scale.

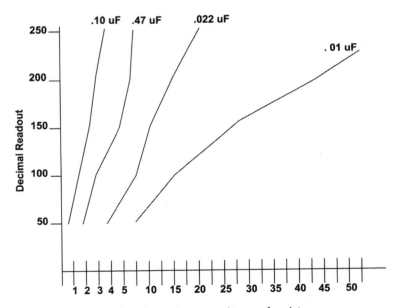

FIGURE 14-2   Graph of numeric readout for various capacitors and resistances

Ideally, with the proper value capacitor and scale, the minimum resistance will output a numeric value close to 0, and maximum resistance will output a numeric output close to 255.

PIN Exceptions

I/O pins that are listed as *Time to Live* (TTL) may be used with the Pot command. Special function pins, such as those listed as *Schmitt triggers* (ST), may not work properly with the Pot command. When using other microcontrollers, remember to check the data sheet for special function pins. With the PICBasic compiler, you are restricted in using the Pot command on Port B pins only. It just so happens that three of the Port B pins (RB0, RB6, and RB7) are listed as a combination TTL/ST. Pins listed as TTL/ST on the 16F84 work with the Pot command.

The data sheet states that RB6 and RB7 are an ST input when used in serial programming mode. RB0 is an ST input when configured as an external interrupt.

Resistive Sensors

Many resistive-type transducers may be read using the Pot command. The important thing to remember is that the Pot command is not an *analog-to-digital* (A/D) converter. Converters measure analog voltages, not resistance.

This may be confusing at first, because A/D converters can read a voltage drop across a resistor and that reading of a voltage drop across a resistor may appear similar to a Pot diagram. It is not! To determine the difference, the key point to look for is a voltage (or current) source on top of the resistive component. If a voltage source exists, you must use an A/D converter to read the output. On the other hand, if the transducer (resistive component) is a passive resistive device within the 50,000-ohm parameters, you may use the Pot command. A/D converters will be examined in the next chapter.

Test Program

Let's work with the Pot command and see how it functions. The first resistive-type sensor you will look at is a Flex Sensor (see Figure 14-3).

Flex sensors have numerous applications. For instance, they may be used as robotic whiskers, which serve as bump and wall sensors. They have been used to create virtual-reality data gloves and biometric measurements for exercise, rehab, and physics applications.

The program uses the *liquid crystal display* (LCD) from Chapter 13, "Liquid Crystal Display (LCD)," to provide a visual readout. The numeric readout, *with the sensor at its maximum resistance,* provides the proper scale factor that one should use in the command to achieve the greatest range with this particular sensor.

For my test, I plugged the Flex sensor into the prototyping breadboard (see Figure 14-4). The schematic for the project is shown in Figure 14-5. Record the numeric readout

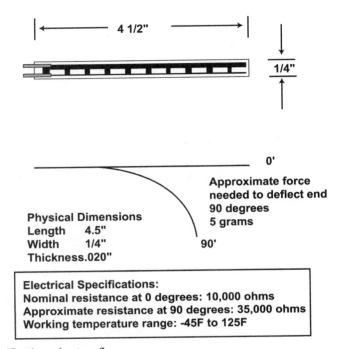

Physical Dimensions
Length   4.5"
Width   1/4"
Thickness.020"

0'

Approximate force
needed to deflect end
90 degrees
5 grams

90'

Electrical Specifications:
Nominal resistance at 0 degrees: 10,000 ohms
Approximate resistance at 90 degrees: 35,000 ohms
Working temperature range: -45F to 125F

FIGURE 14-3   Specifications sheet on flex sensor

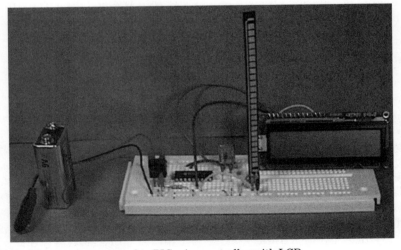

FIGURE 14-4   Flex sensor connected to PIC microcontroller with LCD

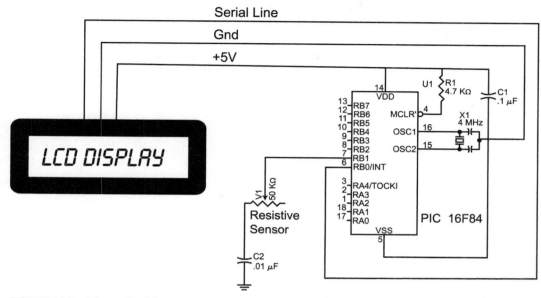

FIGURE 14-5   Schematic of flex sensor and LCD connected to PIC microcontroller

when the sensor is at its maximum resistance, and use that as a scale factor for the best range and accuracy.

```
'PICBasic Program 12.1
'Find Scale Factor and/or Read the Resistive Sensor
Start:
Pot 1,255,B0               'Read resistance on Pin 1 to
                           'determine scale.
Serout 0,N2400,(254,1)     'Clear LCD Screen
Serout 0,N2400, (#B0)      'Send pot values out on pin 0 serially.
Pause 500                  'Wait 1/2 second
Goto start                 'Do it again
```

Here's the PICBasic Pro version:

```
'PICBasic Pro Program 12.2
'Find Scale Factor and/or Read the Resistive Sensor
B0 var byte
Start:
Pot 1,255,B0               'Read resistance on Pin 1 to
                           'determine scale.
Serout 0,4, [254,1]        'Clear LCD Screen
Serout 0,4, [#B0]          'Send pot values out on pin 0 serially
Pause 500                  'Wait 1/2 second
Goto start                 'Do it again
```

If the scale factor displayed by program 12.1 or 12.2 is 255 (or close to 255), the program is already reading the sensor at the best scale (with that particular capacitor).

The LCD is read to determine that the program and sensor are both working properly. The microcontroller of course does not require an LCD to read the sensor. It can read the numeric value held in the variable and "interpret" the results. The LCD is for your benefit.

Fuzzy Logic and Neural Sensors

The Pot command presents you with a few interesting possibilities regarding the interpretation of sensor readings. Here you can have the microcontroller mimic the function of a neural and/or fuzzy logic device.

Fuzzy First

In 1965, Lotfi Zadah, a professor at the University of California at Berkley first published a paper on fuzzy logic. Since its inception, fuzzy logic has been both hyped and criticized.

In essence, fuzzy logic attempts to mimic the way people apply logic in grouping and feature determination. A few examples should clear this "fuzzy" definition. No universal thermometer states at 81.9 degrees Fahrenheit it is a warm day and at 82 degrees Fahrenheit it is a hot day. So when is a warm, sunny day determined not to be a warm day but a hot day instead?

When someone considers a warm day not to be warm but hot depends on that person's personal heat threshold and the influence of their environment and season (see Figure 14-6).

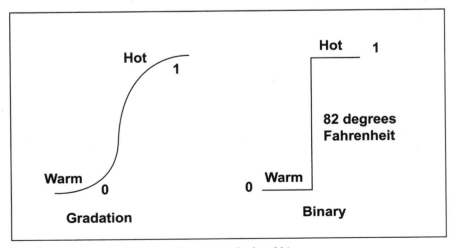

FIGURE 14-6   Two graphs of temperature change, gradual and binary

Examining this example further, a person living in Alaska has a different set of temperature values for hot days when compared to a person living in New York, and both these values will be different from someone living in Florida. And let's not forget seasonal variations. A hot day has a different temperature scale in winter than in summer. So what this boils down to is that fuzzy logic is based on the consensus of many to create a classification. In this particular case, it is the question of what is a hot day and then defining that classification as a group.

Any particular temperature holds a membership in that group. The membership's value (or strength) is determined by how closely that temperature matches the median value.

The group classification theme may be applied to many other things, such as navigation, speed, and height. Let's use height for one more example. If you graph the height of 1,000 people, your graph will resemble the first graph shown in Figure 14-7. You can use this graph of heights to classify shortness, the average height (median), and tallness. If you apply a hard rule that states everyone under 5'7" is short, everyone taller than 6'0" is tall, and everyone in between is medium, your graph will resemble the second graph in Figure 14-7.

Where this hard classification fails is that it classifies a person who is 5'11.5" inches tall as medium. In actuality, the person's height really puts them much more closer to the tall (6' 0" and over) group than the median group.

Height

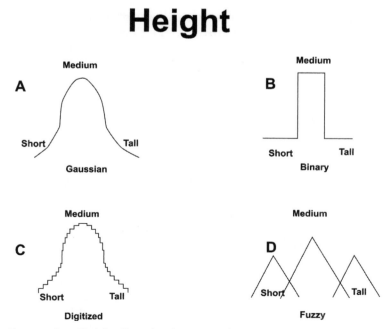

FIGURE 14-7 Four graphs of height: Gaussian, binary, digitized, and fuzzy

This is what fuzzy logic does; instead of hard rules, it mimics people's grouping of rules that's soft and imprecise (or fuzzy). Fuzzy logic uses groups and quantifies the memberships in that group. Groups overlap, as shown in the fourth graph in Figure 14-7. The person who is 5 feet 11.5 inches tall is almost out of the medium group (small membership) and well into the tall group (large membership).

Fuzzy logic provides an alternative to the digitized graph shown. A high-resolution digitized graph is also accurate in classifying height. Why would one choose a fuzzy logic method over a digitized model function? Simplified mathematics and learning functions.

To implement a primitive fuzzy logic in a PIC microcontroller, we will assign a numeric range to a group. This is what you will do in your next project. You will not gauge membership within the group.

Fuzzy Logic Light Tracker

The next project you will build is a fuzzy logic light tracker. The tracker follows a light source using fuzzy logic.

The sensor needed for the tracker is a *cadmium sulfide* (CdS) photocell. These photocells are light-sensitive resistors (see Figure 14-8). Their resistance varies in proportion to the light intensity falling on its surface. In complete darkness, the cell produces its greatest resistance.

Many types of CdS cells are on the market. Your choice of a particular cell will be based on its dark resistance and light-saturation resistance. The term light saturation refers to the state where increasing the light intensity to the CdS will not decrease its resistance any further; it is saturated. The CdS cell I used has approximately a 100K resistance in complete darkness and a 500-ohms resistance when totally saturated with light. Under ambient light, the resistance varies between 2.5 and 10 kilo-ohms.

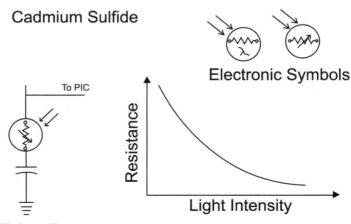

FIGURE 14-8   CdS photocell

To ascertain the proper scale factor, you must first decide which capacitor you should use for the best overall range. Connect the sensor to your PIC microcontroller and run the scale program. Use the scale factor in the program.

The project requires two CdS cells. Test each cell separately. There may be an in-group variance that may change the scale factor used in each cell. In this project, I used a .022 *microfarad* (uF) capacitor, with the scale parameter set at 255 for both cells in the Pot command.

The schematic is shown in Figure 14-9. The CdS cells are connected to Port B pins 2 and 3 (physical pin numbers 8 and 9). The photocells are mounted on a small piece of wood or plastic (see Figure 14-10). Two small holes are drilled for each CdS cell for the wire leads to pass through. Longer wires are soldered onto these wires and connected to the PIC microcontroller.

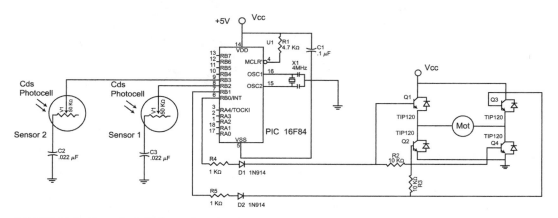

FIGURE 14-9 Fuzzy logic light tracker circuit

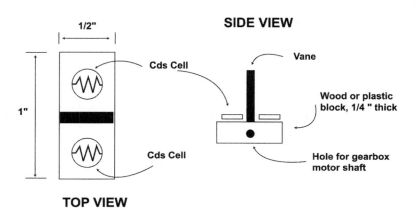

TOP VIEW

FIGURE 14-10 Sensor array

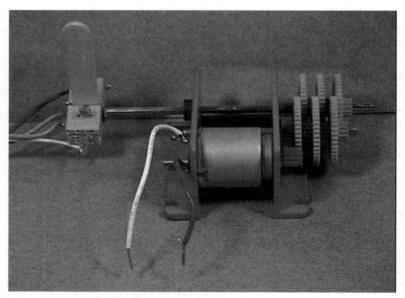

FIGURE 14-11   Sensor array connected to shaft of gearbox motor

One 3/32- to 1/8-inch hole is drilled for the gearbox motor's shaft, and the sensor array is glued to the gearbox motor shaft (see Figure 14-11).

The operation of the tracker is shown in Figure 14-12. When both sensors are equally illuminated, their respective resistance is approximately the same. As long as each sensor is within plus or minus 10 points of the other, the PIC program sees them as equal and doesn't initiate movement. This provides a group range of 20 points. This group range is the fuzzy part in the fuzzy logic.

When either sensor falls in a shadow, its resistance increases beyond your range (group), and the PIC microcontroller activates the motor to bring both sensors under even illumination.

DC Motor Control

The sun tracker uses a gearbox motor to rotate the sensor array toward the light source (see Figure 14-13). The gearbox motor shown has a 4,000:1 ratio. The shaft spins at approximately 1 RPM. You need a suitable slow motor (gearbox) to turn the sensory array, which is attached (glued) to the shaft of the gearbox motor. The gearbox motor can rotate the sensor array *clockwise* (CW) and *counterclockwise* (CCW), depending upon the direction of current flowing through the motor.

"A" cell in shadow tracker rotates to right.

Equal illumination; no movement.

"B" cell in shadow tracker rotates to left.

SIDE VIEW

FIGURE 14-12   Functional behavior of sensor array

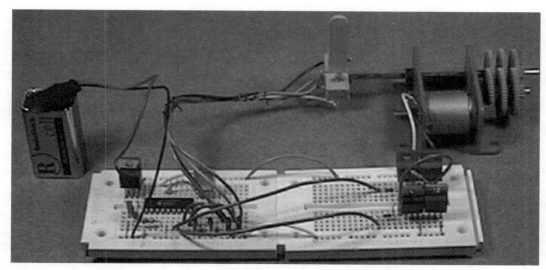

FIGURE 14-13   Fuzzy light tracker circuit

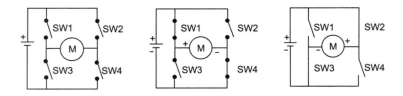

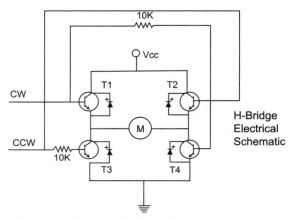

FIGURE 14-14   H-Bridge function and electrical schematic

To rotate the shaft (and sensor array) CW and CCW, you need a way to reverse the current going to the motor, and you will use what is known as an H-Bridge. An H-Bridge uses four transistors (see Figure 14-14). Consider each transistor as a simple on and off switch, as shown in the top portion of the figure. It's called an H-Bridge because the transistors (switches) are arranged in an H pattern. DC motor control and H-Bridges are explained in greater detail in Chapter 16, "DC Motor Control."

When switches SW1 and SW4 are closed, the motor rotates in one direction. When switches SW2 and SW3 are closed, the motor rotates in the opposite direction. When the switches are opened, the motor is stopped.

The PIC microcontroller controls the H-Bridge made of four TIP120 Darlington *negative-positive-negative* (NPN) transistors, four diodes (incorporated inside the TIP120 transistors), and two 10 kilo-ohm, 1/4-watt resistors. Pin 0 is connected to transistors Q1 and Q4, whereas Pin 1 is connected to transistors Q3 and Q4. Using either Pin 0 or 1, the proper transistors are turned on and off to achieve CW or CCW rotation. The microcontroller can stop, rotate CW, or rotate CCW, depending on the reading from the sensor array. Make sure the 10K resistors are placed properly or the H-Bridge will not function.

The TIP120 Darlington transistors are shown in the figure as standard NPN transistors with diodes across the collector-emitter. Many H-Bridge circuit designs use *positive-negative-positive* (PNP) transistors on the high side of the H-Bridge. The on resistance of PNP transistors is higher than NPN transistors, so by using NPN transistors exclusively in your H-Bridge you achieve a little higher efficiency.

Diodes

The diodes shown in the schematic are not components you need to add to the circuit. The diodes are actually built into the TIP120 Darlington transistors. The PIC is sensitive to electrical spikes in its power supply. Electrical spikes could cause the microcontroller to reset or lock up. The diodes across the collector-emitter junction of each transistor (Q1 to Q4) snub any electrical spikes caused by switching the motor's windings on and off. The following is the code for the light tracker:

```
'PICBasic Program 14.3
'Fuzzy Logic Light Tracker
start:
low 0                          'Pin 0 low
low 1                          'Pin 1 low
pot 2,255,b0                   'Read 1st CdS Sensor
pot 3,255,b1                   'Read 2nd CdS Sensor
if b0 = b1 then start          'If equal, do nothing
if b0 > b1 then greater        'If greater, check how much greater
if b0 < b1 then lesser         'If lesser, check how much lesser
greater:                       'Greater routine
b2 = b0 - b1                   'Find the difference
if b2 > 10 then cw             'Is it within range? If not, go to cw
goto start                     'In range do again
lesser:                        'Lesser routine
b2 = b1 - b0                   'Find the difference
if b2 > 10 then ccw            'Is it within range? If not, go to ccw
goto start                     'Do again
cw:                            'Turn the sensor array clockwise
high 0                         'Turn on h-bridge
pause 100                      'Let it turn for a moment
goto start                     'Check again
ccw:                           'Turn the sensor array counter clockwise
high 1                         'Turn on h-bridge
pause 100                      'Let it turn a moment
goto start                     'Check again
```

The PICBasic Pro version is as follows:

```
'PICBasic Pro Program 14.4
'Fuzzy Logic Light Tracker
b0 var byte
b1 var byte
b2 var byte
start:
low 0                          'Pin 0 low
low 1                          'Pin 1 low
```

```
pot 2,255,b0              'Read 1st CdS Sensor
pot 3,255,b1              'Read 2nd CdS Sensor
if b0 = b1 then start     'If equal, do nothing
if b0 > b1 then greater   'If greater, check how much greater
if b0 < b1 then lesser    'If lesser, check how much lesser
greater:                  'Greater routine
b2 = b0 - b1              'Find the difference
if b2 > 10 then cw        'Is it within range? If not, go to cw
goto start                'In range do again
lesser:                   'Lesser routine
b2 = b1 - b0              'Find the difference
if b2 > 10 then ccw       'Is it within range? If not, go to ccw
goto start                'Do again
cw:                       'Turn the sensor array clockwise
high 0                    'Turn on h-bridge
pause 100                 'Let it turn for a moment
goto start                'Check again
ccw:                      'Turn the sensor array counter clockwise
high 1                    'Turn on h-bridge
pause 100                 'Let it turn a moment
goto start                'Check again
```

Operation

When run, the light tracker will follow a light source. If both CdS cells are approximately evenly illuminated, the tracker does nothing. To test it, cover one CdS sensor with your finger. This should activate the gearbox motor, and the shaft should begin to rotate.

If the shaft rotates in the opposite direction of the light source, reverse *either* the sensor input pins or the output pins to the H-Bridge, *but not both*.

Fuzzy Output

The output of your fuzzy light tracker is binary. The motor is either on or off, rotating CW or CCW. In many cases, you would want the output to be fuzzy also. For instance, let's say you're making a fuzzy controller for elevators. You would want the elevator to start and stop gradually (fuzzy), not abruptly as in binary (on or off).

Could you change the output of your light tracker and make it fuzzy? Yes. Instead of simply switching the motor on, you could feed a *pulse width modulation* (PWM) signal that can vary the motor's speed.

Ideally, the motor's speed would be in proportion to the difference (in resistance) of the two CdS cells. A large difference would produce a faster speed than a small difference. The motor speed would change dynamically (in real time) as the tracker brings both CdS cells to equal illumination. This output program may be illustrated using fuzzy logic graphics, groups, and membership sets.

In this particular application, creating a fuzzy output for this demonstration light tracker unit is overkill. If you want to experiment, begin by using the PULSOUT and PWM commands to vary the DC motor speed.

Neural Sensors (Logic)

With a small amount of programming, you can change your fuzzy logic sensors (CdS photocells) to neural sensors. Neural networks are an expansive topic, but this section will be limited to one small example. For those who want to pursue further study into neural networks, I recommend a book I've written titled *Understanding Neural Networks* (ISBN #0-7906-1115-5).

To create a basic neural sensor, take the numeric resistive reading from each sensor, multiply it by a weight factor, and then sum the results. The results are then compared to a tri-level threshold value (see Figure 14-15).

Thus, your small program and sensors are performing all the functions one expects to find in a neural network. You may even be pioneering an original neural event by applying a multivalue threshold scheme.

Are multivalue thresholds natural or mimic-like in nature (biological systems)? The answer is yes, they are both. For instance, an itch is a extremely low level of pain, whereas the sensation of burning is actually the combination of sensing ice cold with warm. The neural sensors you program operate in the same way.

Multivalue Threshold

Typically in neural networks, individual neurons have a singular threshold (positive or negative), and once the threshold is exceeded, it activates the output of the neuron. In

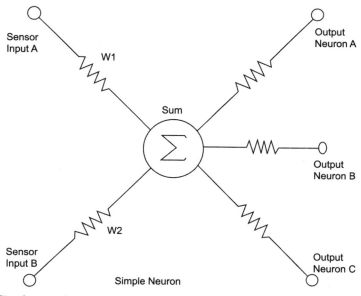

FIGURE 14-15 Simple neural sensor

your example, the output is compared to multivalues, with the output going to the best fit.

Instead of thinking of the output as a numeric value, think of each numeric range as a shape instead; a circle, square, or triangle will suffice. When the neuron is summed, it outputs a shape block (instead of a number). The receptor neurons (*light-emitting diodes* [LEDs]) have a shaped receiver unit that can fit in a shape block. When a shape block matches the receiver unit, the neuron becomes active (the LED turns on).

In this case, each output neuron relates to a particular behavior, such as sleeping, hunting, or feeding, which would be essential behavior for survival in a photovore-style robot (a light eater). Each output represents the different light levels:

- **Low-light level** The photovore stops hunting and searching for food (light). It enters a sleep or hibernation mode.

- **Medium-light level** The photovore hunts and searches for the brightest light areas.

- **High-light level** The photovore stops to feed via solar cells to recharge its batteries.

Instead of building a photovore robotic, you will use an LED to distinguish between each behavior state (see Figure 14-16). Label the LEDs sleeping, hunting, and feeding. Each LED will become active depending on the light level received by the CdS cells. The finished project is shown in Figure 14-17.

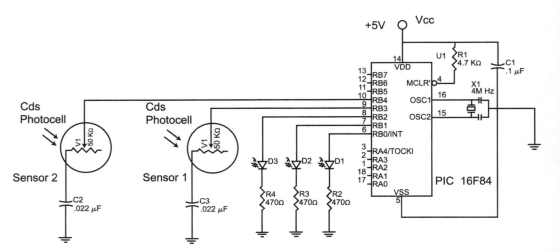

FIGURE 14-16 Neural microcontroller circuit schematic

FIGURE 14-17 Neural microcontroller circuit

The code for the neural microcontroller is as follows:

```
PICBasic Program 14.5 Neural Demo
'Set up
low 0                        'LED 0 off "Sleep"
low 1                        'LED 1 off "Hunt"
low 2                        'LED 2 off "Feed"
Start:
pot 3,255,b0                 'Read 1st sensor
pot 4,255,b1                 'Read 2nd Sensor
w2 = b0 * 3                  'Apply weight
w3 = b1 * 2                  'Apply weight
w4 = w2 + w3                 'Sum results
'Apply thresholds
if w4 < 40 then feed         'Lots of light feed
if w4 <= 300 then hunt       'Medium light hunt
if w4 > 300 then snooze      'Little light sleep
'Actions
feed:                        'Feeding
low 0
low 1
high 2
goto start
hunt:                        'Hunting
low 0
high 1
```

```
low 2
goto start
snooze:                  'Sleeping * DON'T USE KEYWORD SLEEP *
high 0
low 1
low 2
goto start
```

The PICBasic Pro version is as follows:

```
PICBasic Pro Program 14.6 Neural Demo
'Set up
b0 var byte
b1 var byte
w2 var word
w3 var word
w4 var word
low 0               'LED 0 off "Sleep"
low 1               'LED 1 off "Hunt"
low 2               'LED 2 off "Feed"
Start:
pot 3,255,b0        'Read 1st sensor
pot 4,255,b1        'Read 2nd Sensor
w2 = b0 * 3         'Apply weight
w3 = b1 * 2         'Apply weight
w4 = w2 + w3        'Sum results
'Apply thresholds
if w4 < 40 then feed 'Lots of light feed
if w4 <= 300 then hunt 'Medium light hunt
if w4 > 300 then snooze 'Little light sleep
'Actions
feed:               'Feeding
low 0
low 1
high 2
goto start
hunt:               'Hunting
low 0
high 1
low 2
goto start
snooze:             'Sleeping * DON'T USE KEYWORD SLEEP *
high 0
low 1
low 2
goto start
```

Parts List

The components outlined in Chapter 1, "Microcontroller," are to be used here along with the parts listed in Table 14-1. Everything is available from the Images Company, James Electronics, JDR Microdevices, and RadioShack (see the suppliers index).

TABLE 14-1 Additional components

2	CdS photocells
1	Flex sensor (nominal resistance 10K)
2	.022 uF capacitors
1	.01 uF capacitor
4	TIP120 NPN Darlington transistors
2	10K resistors
2	1N514 diodes
2	1K resistors
1	Gearbox motor

Analog-to-Digital (A/D) Converters

Analog-to-digital (A/D) converters are specially designed *integrated circuits* (ICs) that can read an analog voltage and convert it into an equivalent digital number. That digital number can be read and interpreted by the microcontroller (or computer) to ascertain a close approximation of the value of the analog voltage being presented at the A/D converter.

The reason this is important is that most natural phenomena, when measured, have an analog value. For instance, when using instruments, sensors, or transducers to measure the intensity of light, sound, temperature, pressure, time, weight, or gravity, analog values will be recorded.

Analog Signal

Any analog signal, regardless of its origin, is infinitely variable between any two measuring points. This holds true no matter how close those two measuring points are. For instance, the number of possible volt readings between 1 volt and 2 volts is infinite. Some possible values are 1.1 volts, 1.0000001 volts, or 1.00000000000000000001 volts. As you can see, voltage can vary by infinitesimal amounts, making the possible values infinite.

Digital Equivalents

A digital equivalent value of an analog signal is not infinitely variable. The digital equivalent can only change in discrete, predefined steps that are determined by the resolution of the A/D converter. Figure 15-1 illustrates an analog signal and a digital equivalent. A rising voltage (analog signal) plotted digitally against time could only jump in increments.

Each bar represents a specific voltage that may jump up or down based on the resolution of the A/D converter. The resolution of the A/D converter is illustrated by the vertical tick marks.

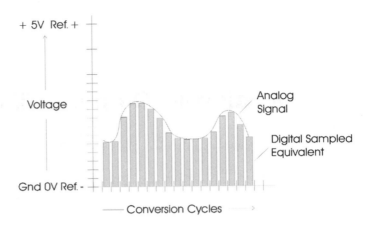

FIGURE 15-1   Plot of analog signal and digital sampled equivalent

You can also observe from the figure that the digital equivalents are only updated once per clock cycle. Complete clock cycles are indicated on the drawing as the horizontal tick marks. In between the sampling clock cycles, no reading (conversions) are performed.

If the analog signal changes significantly and quickly returns to its previous value between digital sampling updates, the A/D converter will miss that signal change (see Figure 15-2).

According to the Nyquist Theorem, you should sample your analog signal at twice the frequency of the highest signal frequency you are attempting to sample. If you want to sample a signal that may be changing at 5,000 Hz, you ought to sample the signal with an A/D converter that is capable of sampling at 10,000 Hz.

For instance, human hearing extends upward to 20,000 Hz. It's not a coincidence that digital sound (CDs) is sampled and recorded at 44,000 Hz. The sample presecond capacity of the A/D converter is indicated on the drawing(s) as horizontal tick marks. A/D converters are typically rated by how many samples per second (conversions) they are capable of performing.

A/D Converters

Certain PIC microcontrollers have built-in A/D converters. Since you have based your work on the PIC 16F84, I will continue to work with this chip by connecting an external A/D converter. (My next book will deal with more advanced PIC microcontrollers with built-in A/D converters.)

To minimize the *input/output* (I/O) lines needed, we will use a serial A/D converter. The TLC549 is shown in Figure 15-3. This serial A/D chip will require just three lines of your PIC microcontroller. The specifications on this A/D converter are as follows:

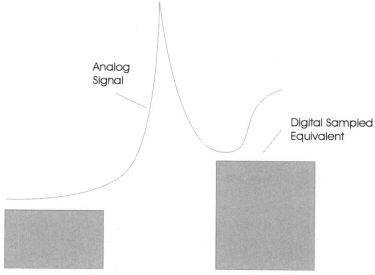

FIGURE 15-2 Missed analog signal change

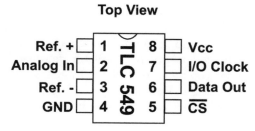

FIGURE 15-3 TLC 549 serial A/D converter

- *Complementary metal oxide semiconductor* (CMOS) technology
- Eight-bit resolution (256)
- Reference input voltages (+/−)
- Conversion time, 17 *microseconds* (µsec) maximum
- 40,000 samples per second (approximately)
- Built-in sample and hold

- Wide power supply range (3 to 6 volts)
- 4 MHz internal clock
- Low-power consumption 6 *milliwatt* (mW) (typical)

This chip is easily interfaced to your microcontroller.

Setting the Reference Voltage(s)

Looking at the IC shown in Figure 15-3, two pins are used for setting the reference voltages: Ref+ (pin 1) and Ref− (pin 3). The voltages placed between these two pins become the range of voltages that the A/D converter will read and convert to a digital equivalent.

The voltage difference between the two reference (Ref) pins must be at least 1 volt. The Ref+ voltage should not be greater than the power supply voltage being supplied to the chip (Vcc). Consequently, the Ref− voltage should not be less than the ground (GND) supply to the chip.

If desired, the Ref+ pin can be tied to Vcc, and Ref− can be tied to GND. This will enable the chip to read voltages between GND and Vcc.

Voltage Range and Resolution

Assuming a Vcc of +5 volts, with Ref+ tied to Vcc and Ref− tied to ground, as shown in Figure 15-1, what is the resolution of your converter chip? Take your voltage range from Ref − to Ref+, in this case 5 volts, and divide by your 8-bit resolution, 256, which equals .019 volts. You would then know that each numerical increment in the output of the A/D converter IC is equal to a +.019-volt increase in voltage.

Suppose the sensor you need to read a voltage from only varies by two volts, say, from 1 to 3 volts. If you wanted to increase the resolution of your A/D converter, you could set Ref− to 1 volt and Ref+ to 3 volts, as shown in Figure 15-4. Now what is the resolution of your A/D converter? It's calculated just as before. Take the voltage range from Ref − to Ref+, in this case 2 volts, and divide by 256 (8-bit resolution): $2/256 = .0078$ volts.

Interpreting the Results

Suppose the PIC microcontroller is reading the number 100 from the serial A/D converter. What does this number represent? Let's go back to the first case, Figure 15-1, where Vcc is 5 volts, the voltage range is 5 volts, Ref+ is tied to Vcc, and Ref− is tied to GND. Your resolution is .019 volts, so reading 100 from the A/D chip means it is reading a voltage of 1.9 volts ($100 \times .019$ volts).

In the second case, Ref− is at 1 volt, Ref+ is at 3 volts (refer to Figure 15-4), the range equals 2 volts, and the step resolution equals .0078 volts. Here reading the number 100 from the serial A/D converter is equal to a voltage of 1.78 volts ($100 \times .0078$ volts = .78 volts; .78 volts plus Ref− [1 volt] = 1.78 volts).

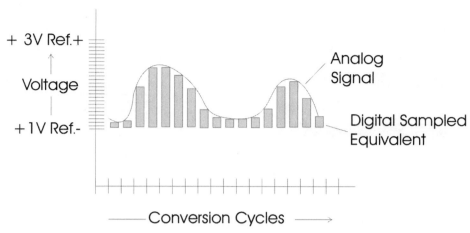

FIGURE 15-4   Plot of analog signal within Ref+ and Ref− points

Serial A/D Converter Chip Control

With the basic calculations finished, it's time to actually implement the chip. You need three I/O lines to use the serial A/D chip. The CS pin stands for Chip Select, and the small line or bar above the CS tells us that the pin is active. So, when the CS pin is brought low, the chip is selected. A clock signal is sent to the chip's I/O clock pin. You will read the serial data from the data out pin.

You will have to create your own serial routine, as you did in Chapter 12, "Creating a New I/O Port," to read the serial A/D IC because the RS-232 communication protocol in the Serial In (Serin) routine will become fouled up due to the stop and start bits. However, you will use the RS-232 output routine (Serout) to display the information from your A/D onto your *liquid crystal display* (LCD).

TLC549 Serial Sequence

This sequence shows how the serial A/D chip can be accessed easily:

1. Bring the CS pin low. This immediately places the *most significant bit* (MSB) on the data out pin.

2. The falling edges of the first four I/O clock cycles shift out the second, third, fourth, and fifth MSBs. At this point, the on-chip sample and hold begin sampling the analog input.

3. Three more I/O clock cycles bring forth the sixth, seventh, and eighth conversion bits on the falling edges.

4. Bring the CS high and the I/O clock line low.

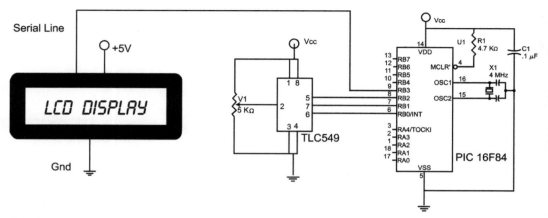

FIGURE 15-5   Test circuit using a 10 kilo-ohm potentiometer

The first schematic is shown in Figure 15-5. The circuit is using a 5K or 10K potentiometer to create a varying voltage source. By moving the potentiometer's wiper back and forth, the voltage seen by the serial A/D converter changes. The digital equivalent number outputted on the LCD changes as the wiper is moved. You can calculate the voltage the number represents by multiplying the number on the display by .019 volts. The code for the program is as follows:

```
'PICBasic Serial A/D Converter Program 15.1
low 1           'Bring I/O clock low
start:
gosub serial_in
' LCD Routine
serout 3, N2400, (254, 1)
pause 1
serout 3, N2400, (#b0)
pause 100       'Let me see display
goto start
'Serial In Routine
serial_in:
low 2           'Bring CS down low
bit7 = pin0     'Load bit7 into B0
pulsout 1,1     'Bring clk pin high then low
bit6 = pin0     'Load bit6 into B0
pulsout 1,1     'Bring clk pin high then low
bit5 = pin0     'Load bit5 into B0
pulsout 1,1     'Bring clk pin high then low
bit4 = pin0     'Load bit4 into B0
pulsout 1,1     'Bring clk pin high then low
bit3 = pin0     'Load bit3 into B0
pulsout 1,1     'Bring clk pin high then low
bit2 = pin0     'Load bit2 into B0
pulsout 1,1     'Bring clk pin high then low
bit1 = pin0     'Load bit1 into B0
```

```
pulsout 1,1       'Bring clk pin high then low
bit0 = pin0       'Load bit0 into B0
pulsout 1,1
high 2            'Bring CS high
return
```

For PICBasic Pro, the program is as follows:

```
'PICBasic Pro Serial A/D Converter Program 15.2
b0 var byte
low 1             'Bring I/O clock low
start:
gosub serial_in
' LCD routine
serout 3, 4, [254, 1]
pause 1
serout 3, 4, [#b0]
pause 100         'Let me see display
goto start
'Serial In routine
serial_in:
low 2             'Bring CS down low
b0.7 = pin0       'Load bit7 into B0
pulsout 1,1       'Bring clk pin high then low
b0.6 = pin0       'Load bit6 into B0
pulsout 1,1       'Bring clk pin high then low
b0.5 = pin0       'Load bit5 into B0
pulsout 1,1       'Bring clk pin high then low
b0.4 = pin0       'Load bit4 into B0
pulsout 1,1       'Bring clk pin high then low
b0.3 = pin0       'Load bit3 into B0
pulsout 1,1       'Bring clk pin high then low
b0.2 = pin0       'Load bit2 into B0
pulsout 1,1       'Bring clk pin high then low
b0.1 = pin0       'Load bit1 into B0
pulsout 1,1       'Bring clk pin high then low
b0.0 = pin0       'Load bit0 into B0
pulsout 1,1
high 2            'Bring CS high
return
```

Toxic Gas Sensor

The toxic gas sensor responds to a large number of airborne compounds (see Figure 15-6). It is a resistive device, and when it detects airborne compounds, its resistance decreases. Figure 15-7 is a schematic of the toxic gas sensor project. Pins 2 and 5 on the gas sensor are connected to an internal heater coil. The heater draws 115 *milliamperes* (mA) at 5 volts. Pins 4 and 6 are connected together internally, as are Pins 1 and 3. You can solder wires to the sensor directly or purchase a round six-pin socket.

Polarity isn't important to the heater coil or resistive element. You may notice that, as the sensor is operating, it will feel quite warm. Don't be alarmed; this is normal.

Since the sensor has been in storage prior to you receiving it, it will require an initial two-minute warmup when it is first turned on. This warmup period will decrease

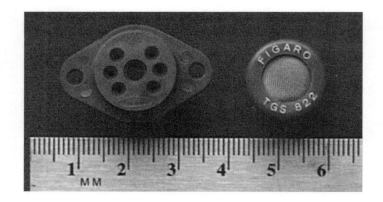

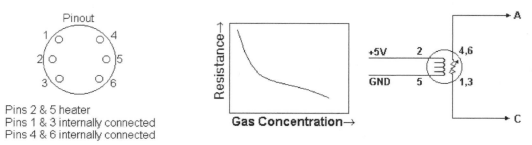

Pins 2 & 5 heater
Pins 1 & 3 internally connected
Pins 4 & 6 internally connected

FIGURE 15-6   Photograph and response of toxic gas sensor

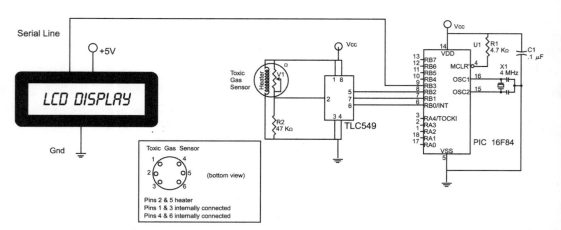

FIGURE 15-7   Schematic using toxic gas sensor

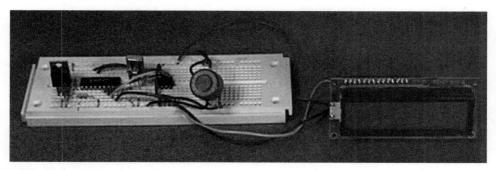

FIGURE 15-8   Toxic gas sensor circuit

TABLE 15-1   Additional components

TLC549 serial A/D	$12.95
Toxic gas sensor	$19.95
Six-pin socket (gas sensor)	$1.50

with repeated use. After the warmup, you can test the sensor with a number of household items.

For the first test, breathe on the sensor. The numbers on the LCD should appear as they respond to the carbon dioxide in your breath. For another test, release gas from a butane lighter; the sensor should react immediately to the butane gas. Figure 15-8 is a photograph of the complete project. In the next chapter we will use this circuit to make a toxic gas alarm and an automatic ventilator control.

Parts List

The parts needed for this project are listed in Chapter 1, "Microcontrollers," as well as in Table 15-1, and they are available from Images SI, Inc. (see the suppliers index).

DC Motor Control

This chapter will look at a few methods of controlling a DC hobby motor. A single pin of the 16F84 PIC microcontroller is limited to a maximum output current of 25 *milliamps* (mA). In most cases, this is too feeble a current to power a DC motor directly. Instead you must use the output of a PIC pin to turn on and off a transistor that can easily control the current needed to run hobby motors. The two common methods you will use incorporate transistors to switch the current on and off.

The Transistor

The transistor of choice used in most of these examples is a TIP120 *negative-positive-negative* (NPN) transistor. The TIP120 is a Darlington transistor at medium power with a 5-ampere maximum current, and it is designed for general-purpose amplification and low-speed switching. The *positive-negative-positive* (PNP) version of the transistor is the TIP125.

First Method

The TIP120 transistor is a simple on-off motor switch (see Figure 16-1). When the PIC pin is brought high, the transistor goes into conduction, thereby turning on the DC motor. The TIP120 transistor has a built-in diode across the collector-emitter junction of the transistor. This protects the transistor from any inductive voltage surge caused by switching the motor off. For added PIC protection, I inserted a signal diode and current-limiting resistor on the output pin. The following program may be used for the PICBasic and PICBasic Pro compilers:

```
'Program 16.1 DC Motor
pause 1000      'Wait a second
high 0          'Turn on DC motor
```

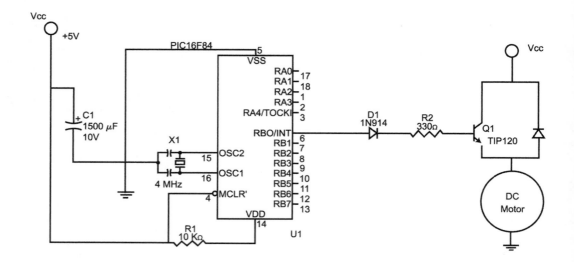

FIGURE 16-1   On-off motor switch using the TIP120 transistor

```
pause 1000      'Wait a second
low 0           'Turn off DC Motor
end
```

In the circuit, notice the 1,500-microfarad (uF) capacitor. A large capacitor is needed to smooth the voltage dips caused by the DC motor turning on and off. Without a large capacitor, the sudden dip in voltage may inadvertently reset the PIC microcontroller. A picture of this circuit is shown in Figure 16-2.

Bidirectional

An H-Bridge allows bidirectional control of a DC motor. To achieve this, it uses four transistors (see Figure 16-3). Consider each transistor as a simple on and off switch, as shown in the top portion of the drawing. The reason this circuit is called an H-Bridge is because the transistors (switches) are arranged in an H-type pattern.

When switches SW1 and SW4 are closed, the motor rotates in one direction. When switches SW2 and SW3 are closed, the motor rotates in the opposite direction. When the switches are opened, the motor is stopped. Replace the switches with transistors and you have an electronic H-Bridge.

The PIC microcontroller controls the H-Bridge made of four TIP120 Darlington NPN transistors. The four diodes across the collector and emitter of each TIP120 transistor

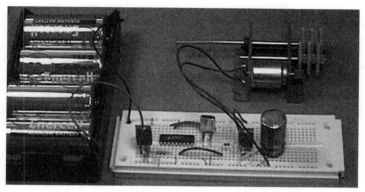

FIGURE 16-2 The built circuit from Figure 16-1

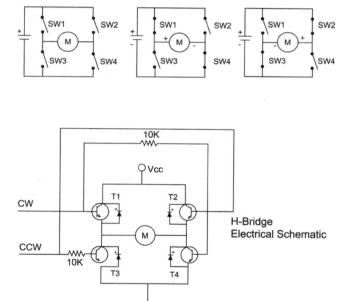

FIGURE 16-3 H-Bridge schematic and function

are internal diodes (see Figure 16-4). The TIP120 Darlington transistors are drawn in the schematic as standard NPN transistors with diodes. Pin 0 is connected to transistors Q1 and Q4, and Pin 1 is connected to transistors Q3 and Q4. Using either Pin 0 or 1,

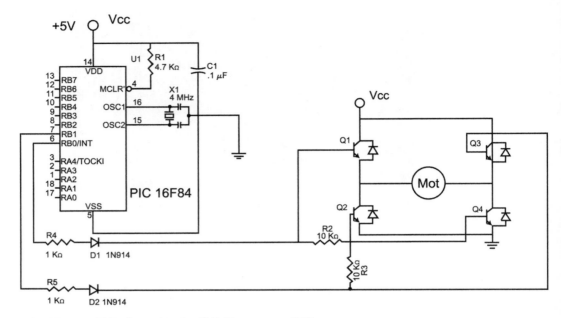

FIGURE 16-4   PIC schematic using H-Bridge to control DC motor

the proper transistors are turned on and off to achieve *clockwise* (CW) or *counterclockwise* (CCW) rotation. If, by accident or programming, error pins 0 and 1 are brought high simultaneously, this will create a short circuit. Using the H-Bridge properly, the microcontroller can stop, rotate CW, or rotate CCW the DC motor.

Many H-Bridge circuit designs use PNP transistors on the high side of the H-Bridge. The on resistance of PNP transistors is higher than that of NPN transistors, so in using NPN transistors exclusively in your H-Bridge you achieve a little better efficiency.

Diodes

PIC microcontrollers are sensitive to electrical spikes (which may cause a reset or lockup). When using standard transistors, you should place diodes across the collector-emitter junction of each transistor. These diodes snub any electrical spikes caused by switching the motor's windings on and off.

The completed project is shown in Figure 16-5. The following program rotates the motor CW for 1 second, pauses for .5 seconds, rotates CCW for 1 second, pauses for .5 seconds, and then repeats the sequence.

```
'H-Bridge Program 16.2 PICBasic and PICBasic Pro
low 0
low 1
```

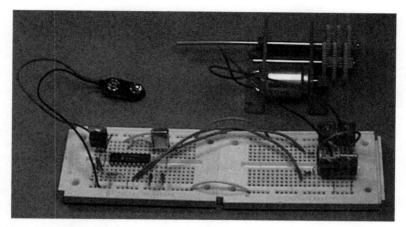

FIGURE 16-5   H-bridge project

TABLE 16-1   Additional components

Quantity	Item
4	TIP120 NPN Darlington transistors
2	10K resistors
2	1K resistors
1	DC motor
2	1N914 signal diodes

```
start:
pause 500      'Pause for .5 sec
high 1         'Rotate motor in one direction
pause 1000     'for 1 second
low 1          'Stop motor
pause 500      'Pause for .5 sec
high 0         'Rotate motor in opposite direction
pause 1000     'for 1 seconds
low 0          'Stop Motor
goto start
```

Parts List

The components from Chapter 1, "Microcontrollers," should be used here along with the parts listed in Table 16-1. They are available from the Images Company, James Electronics, JDR Microdevices, and RadioShack (see the suppliers index).

Stepper Motors

In the last chapter, you programmed the PIC microcontroller to control DC motors. Aside from DC motors, stepper motors and servomotors are two popular motors you can control using PIC microcontrollers. This chapter will examine stepper motors, and the following chapter will focus on servomotors.

Stepper motors provide considerable advantages over DC motors. They allow for precise positioning in a wide range of applications that include robotics, automation, and animatronics.

Stepper motors operate differently from DC motors. When power is applied to a DC motor, the rotor begins turning smoothly. Speed is measured in *revolutions per minute* (RPM) and is a function of voltage, current, and load on the motor. The precise positioning of the motor's rotor is not usually possible or desirable.

A stepper motor, on the other hand, runs on a controlled sequence of electric pulses to the windings of the motor. Each pulse rotates the stepper motor's rotor by a precise increment. Each increment of the rotor is referred to as a step, hence the name stepper motors.

The incremental steps of the rotor's rotation translate to a high degree of positioning control, either rotationally or linearly (if the stepper motor is configured to produce linear motion). The incremental rotation is measured in degrees.

Stepper motors are manufactured with varying degrees of rotation per step. You will find the degrees per step in the specifications of the stepper motor. The degrees per step of many stepper motors vary from a fraction of a degree (.12 degree) to many degrees (that is, 22.5 degrees).

To become familiar with stepper motors, you will build a simple stepper motor controller from a PIC16F84 and examine the operating principals.

Stepper Motor Construction and Operation

Stepper motors are constructed using strong, permanent magnets and electromagnets. The permanent magnets are located on the rotating shaft called the rotor. The

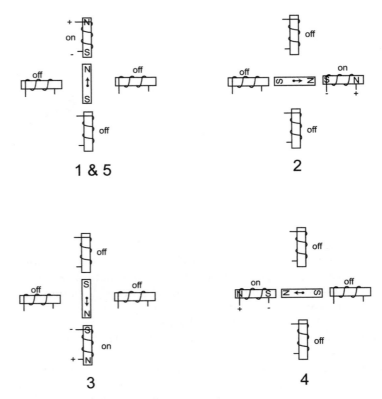

FIGURE 17-1   Stepper motor going through one rotation

electromagnets or windings are located on the stationary portion of the motor called the stator. Figure 17-1 illustrates a stepper motor stepping through one complete rotation. The stator, or stationary portion of the motor, surrounds the rotor.

In Figure 17-1, at position 1, you start with the rotor facing the upper electromagnet that is turned on. To move in a *clockwise* (CW) rotation, the upper electromagnet is switched off, as the electromagnetic to the right is switched on. This causes the rotor to rotate 90 degrees to align itself to the electromagnet in a CW rotation, shown in position 2. Continuing in the same manner, the rotor is stepped through a full rotation until you end up in the same position as you started in position 5.

Resolution

The degree of rotation per pulse is the resolution of the stepper motor. In Figure 17-1, the rotor turned 90 degrees per pulse—not a very practical motor. A practical stepper motor has a greater resolution (smaller steps), such as one that rotates its shaft 1 degree per pulse (or step). This motor requires 360 pulses (or steps) to complete one revolution. When a stepper motor is used for positioning in a linear motion table, each step of the motor translates to a precise increment of linear movement.

Assume that one revolution of the motor is equal to 1 inch of linear travel on the table. For a stepper motor that rotates 3.75 degrees per step, the increment of linear movement is approximately .01 inches per step. A stepper motor that rotates 1.0 degrees per step would give approximately .0027 inches per step. The increment of linear movement is proportional to the degrees per step.

Half-Step

It is possible to double the resolution of some stepper motors by a process known as half-stepping. The process is illustrated in Figure 17-2. In position I, the motor starts with the upper electromagnetic switched on, as before. In position II, the electromagnetic to the right is switched on while keeping power to the upper coil on. Since both coils are on, the rotor is equally attracted to both electromagnets and positions itself between both positions (a half-step). In position III, the upper electromagnet is switched off and the rotor completes one step. Although I am only showing one half-step, the motor can be half-stepped through the entire rotation.

Other Types of Stepper Motors

There are four wire stepper motors. These stepper motors are called bipolar and have two coils, with a pair of leads to each coil. Although the circuitry of the wire stepper motor is simpler than the one you are using, it requires a more complex driving circuit. The circuit must be able to reverse the current flow in the coils after it steps.

Real World

The stepper motor illustrated rotates 90 degrees per step. Real-world stepper motors employ a series of minipoles on the stator and rotor. The minipoles reduce the degrees

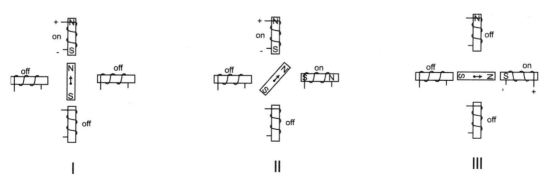

FIGURE 17-2 Half-stepping

per step and improve the resolution of the stepper motor. Although Figure 17-3 appears more complex, its operation is identical to the previous illustrations shown in Figures 17-1 and 17-2.

The rotor in Figure 17-3 turns in a CW rotation. In position I, the north pole of the permanent magnet on the rotor is aligned with the south pole of the electromagnet on the stator. Notice that the multiple positions are all lined up. In position II, the electromagnet is switched off, and the coil to its immediate left is switched on. This causes the rotor to rotate CW by a precise amount. It continues in this same manner for all the steps. After eight steps, the sequence of the electric pulse would start to repeat. Half-stepping with the multipole position is identical to the half-step described previously.

First Stepper Circuit

Figure 17-4 is the schematic for your first test circuit. The output lines from the PIC 16F84 are buffered using a 4050 hex buffer chip. Each buffered signal line is connected to a *negative-positive-negative* (NPN) transistor. The TIP120 transistor is actually an NPN Darlington; in the schematic it is shown as a standard NPN. The TIP120 transistors act like switches, turning on one stepper motor coil at a time.

The diodes shown across each transistor collector and emitter are internal and protect the transistor from the inductive surge created when switching the current on and off in the stepper motor coils. The diode provides a safe return path for the reverse current. Without the internal diodes, the transistor will be more prone to failure and/or a shorter life.

Multipole Operation

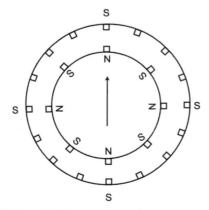
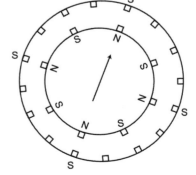

FIGURE 17-3   High-resolution stepper motor

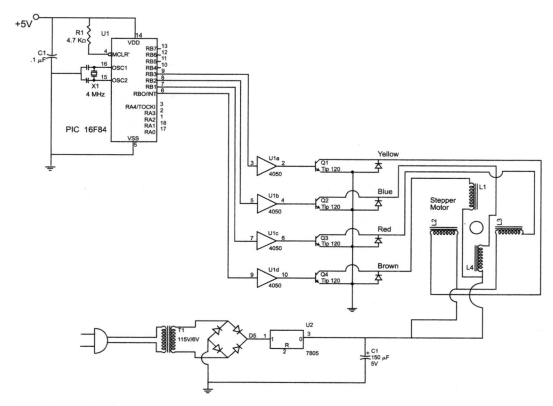

FIGURE 17-4 Schematic for test circuit

Electrical Equivalent of a Stepper Motor

Figure 17-5 is an equivalent electric circuit of the stepper motor you are using. The stepper motor has six wires coming out of the casing. You can see by following the lines that three leads go to each half of the coil windings and that the coil windings are connected in pairs. This is how unipolar, four-phase stepper motors are wired.

Let's assume you just picked this stepper motor and didn't know anything about it. The simplest way to analyze it is to check the electrical resistance between the leads. By making a table of the resistances measured between the leads, you'll quickly find which wires are connected to which coils.

Figure 17-6 shows how the resistance of the motor you are using looks. A 13-ohm resistance exists between the center tap wire and each end lead, with 26 ohms between the two end leads. The resistance reading from the wires originating from the separate

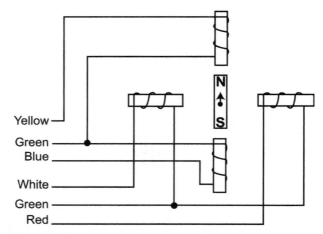

FIGURE 17-5   Electrical equivalent of stepper motor

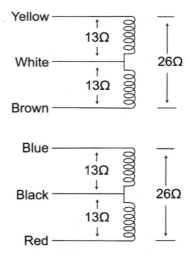

FIGURE 17-6   Testing resistance of stepper motor coils

coils will show an infinitely high resistance (no connection). This would be the case when reading the resistance between the blue and brown leads, for instance.

Armed with this information, you can just about decipher any six-wire stepper motor you come across and wire it properly into a circuit. The stepper motor you are using rotates 1.8 degrees per step.

Test Circuit Program

After you are finished constructing the test circuit, program the PIC with the following basic program. The program is kept small and simple in order to show how easy it is to get a stepper motor moving. Table 17-1 shows that each step in the sequence turns on one transistor. Use the table to follow the logic in the PICBasic program. When you reach the end of the table, the sequence repeats, starting back at the top of the table.

```
'PICBasic Stepper Motor Controller Program 17.1
Symbol TRISB = 134      'Initialize TRISB to 134
Symbol PortB = 6        'Initialize Port B to 6
symbol ti = b6          'Initial ti delay
ti = 25                 'Set delay to 25 ms
Poke TRISB,0            'Set Port B lines output
start:'Forward rotation sequence
poke portb,1            'Step 1
pause ti                'Delay
poke portb,2            'Step 2
pause ti                'Delay
poke portb,4            'Step 3
pause ti                'Delay
poke portb,8            'Step 4
pause ti                'Delay
goto start              'Do again
```

The PICBasic Pro program is as follows:

```
'PICBasic Pro Stepper Motor Controller Program 17.2
ti var byte      'Initial time delay
ti = 25          'Set delay to 25 ms
TRISB = 0        'Set Port B lines output
start:'Forward rotation sequence
portb = 1        'Step 1
pause ti         'Delay
portb = 2        'Step 2
pause ti         'Delay
portb = 4        'Step 3
pause ti         'Delay
portb = 8        'Step 4
pause ti         'Delay
goto start       'Do again
```

TABLE 17-1   Stepper motor sequence

	Full-step transistors			
Q1	Q2	Q3	Q4	Port B output (decimal)
on	—	—	—	1
—	on	—	—	2
—	—	on	—	4
—	—	—	on	8

One Rotation

Using whole steps, the stepper motor requires 200 pulses to complete a single rotation (360 degrees/1.8 degrees per step). Having the PIC microcontroller count the pulses allows it to control and position the stepper motor's rotor.

Second Basic Program

This second PICBasic program is far more versatile. The user can modify programmed parameters (time delay) as the program is running using one of the four switches connected to Port A. Pressing SW1 lengthens the delay pause between steps in the sequence and subsequently makes the stepper motor rotate slower. Pressing SW2 has the opposite effect. By pressing SW3, the program halts the stepper motor and stays in a holding loop for as long as SW3 is closed (or pressed). Rotation direction, either CW or *counterclockwise* (CCW), is controlled with the SW4 switch. Pressing the SW4 switch reverses the stepper motor direction. The direction stays in reverse for as long as SW4 is pressed (or closed). The program is as follows:

```
'PICBasic Stepper Motor Controller Program 17.3
Symbol TRISB = 134        'Initialize TRISB to 134
Symbol TRISA = 133        'Initialize TRISA to 133
Symbol PortB = 6          'Initialize Port B to 6
Symbol PortA = 5          'Initialize Port A to 5
symbol ti = b6            'Initial ti delay
ti = 100                  'Set delay to 100 ms
Poke TRISB,0              'Set Port B lines output
start:'Forward stepper motor rotation sequence
poke portb,1              'Step 1
pause ti                  'Delay
poke portb,2              'Step 2
pause ti                  'Delay
poke portb,4              'Step 3
pause ti                  'Delay
poke portb,8              'Step 4
pause ti                  'Delay
goto check               'Jump to check switch status
start2:                   'Reverse motor rotation sequence
poke portb,8              'Step 1
pause ti                  'Delay
poke portb,4              'Step 2
pause ti                  'Delay
poke portb,2              'Step 3
pause ti                  'Delay
poke portb,1              'Step 4
pause ti                  'Delay
goto check               'Jump to check switch status
Check:'Switch status
Peek PortA, B0            'Peek the switches
If bit0 = 0 then loop1    'If SW1 is closed, increase ti
if bit1 = 0 then loop2    'If SW2 is closed, decrease ti
if bit2 = 0 then hold3    'Stop motor
if bit3 = 0 then start    'Go Forward
goto start2              'Go Reverse
```

```
loop1:'Increase Delay
poke portb,0                    'turn off transistors
ti = ti + 5                     'Increase delay by 5 ms
pause 50                        'delay
if ti > 250 then hold1          'limit delay to 250 ms
peek porta,b0                   'check switch status
if bit0 = 0 then loop1          'still increasing delay?
goto check                      'if not jump to main switch status
loop2:'Decrease Delay
poke portb,0                    'Turn off transistors
ti = ti - 5                     'Decrease delay by 5 ms
pause 50                        'Pause a moment
if ti < 20 then hold2           'Limit delay to 20 ms
peek porta,b0                   'check switch status
if bit1 = 0 then loop2          'still decreasing delay?
goto check                      'if not, jump to main switch status
hold1:'limit upper delay
ti = 245                        'to 250 ms
goto loop1                      ' go back
hold2:'limit lower delay
ti = 25                         'to 25 ms
goto loop2                      'go back
hold3:'Stop stepper motor
poke portb,0                    'Turn off transistor
peek porta, b0                  'Check switches
if bit2 = 0 then hold3          'Keep motor off?
goto check                      'if not, go to main switch status check
```

The PICBasic Pro program is as follows:

```
'PICBasic Pro Stepper Motor Controller Program 17.4
ti var byte                     'Initial ti delay
ti = 100                        'Set delay to 100 ms
TRISB = 0                       'Set Port B lines output
start:'Forward stepper motor rotation sequence
portb = 1                       'Step 1
pause ti                        'Delay
portb = 2                       'Step 2
pause ti                        'Delay
portb = 4                       'Step 3
pause ti                        'Delay
portb = 8                       'Step 4
pause ti                        'Delay
goto check                      'Jump to check switch status
start2:                         'Reverse motor rotation sequence
portb = 8                       'Step 1
pause ti                        'Delay
portb = 4                       'Step 2
pause ti                        'Delay
portb = 2                       'Step 3
pause ti                        'Delay
portb = 1                       'Step 4
pause ti                        'Delay
goto check                      'Jump to check switch status
Check:'Switch status
'Peek Port A to check switch status
If PortA.0 = 0 then loop1       'If SW1 is closed, increase ti
if PortA.1 = 0 then loop2       'If SW2 is closed, decrease ti
```

```
if PortA.2 = 0 then hold3      'Stop motor
if PortA.3 = 0 then start      'Go Forward
goto start2                    'Go Reverse
loop1:'Increase Delay
portb = 0                      'turn off transistors
ti = ti + 5                    'Increase delay by 5 ms
pause 50                       'delay
if ti > 250 then hold1         'limit delay to 250 ms
'Peek PortA to check switch status
if PortA.0 = 0 then loop1      'still increasing delay?
goto check                     'if not, jump to main switch status
loop2:'Decrease Delay
portb = 0                      'Turn off transistors
ti = ti - 5                    'Decrease delay by 5 ms
pause 50                       'Pause a moment
if ti < 20 then hold2          'Limit delay to 20 ms
'Peek Port A to check switch status
if PortA.1 = 0 then loop2      'still decreasing delay?
goto check                     'if not, jump to main switch status
hold1:'limit upper delay
ti = 245                       'to 250 ms
goto loop1                     ' go back
hold2:'limit lower delay
ti = 25                        'to 25 ms
goto loop2                     'go back
hold3:'Stop stepper motor
portb = 0                      'Turn off transistor
'Peek PortA to check switches
if PortA.2 = 0 then hold3      'Keep motor off?
goto check                     'if not, go to main switch status check
```

The schematic for this program is shown in Figure 17-7. In the photograph of the circuit in Figure 17-8, the four switches are difficult to make out. They are the four bare wire strips behind the PIC microcontroller. The top side of the bare wire strips are connected to +5-volt through 10K resistors. A wire from each switch is connected to the appropriate pin on Port A. A single wire is connected to ground and is used to close any of the switches by touching the bare wire strip.

Half-Stepping

Half-stepping the motor effectively doubles the resolution. In this instance, it requires 400 pulses to complete one rotation. Table 17-2 shows the switching logic needed in a program. When you reach the end of the table, the sequence repeats, starting back at the top of the table.

The "ti" Delay Variable

The "ti" variable used in each basic program controls a delay pause whose purpose is to slow down the output sequence to Port B. Without the pause, the sequence may run too fast for the stepper motor to respond, causing the stepper motor to malfunction.

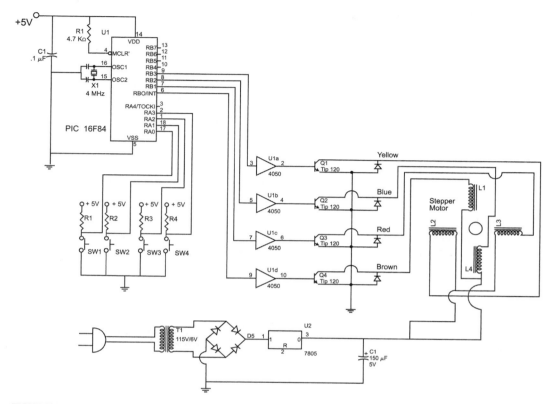

FIGURE 17-7 Schematic of the stepper motor circuit

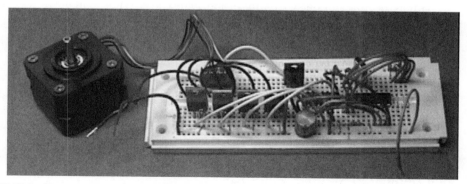

FIGURE 17-8 Stepper motor circuit of Figure 17-7

TABLE 17-2   Program switching logic

	Half-step			
Q1	Q2	Q3	Q4	Port B output (decimal)
on	—	—	—	1
on	on	—	—	3
—	on	—	—	2
—	on	on	—	6
—	—	on	—	4
—	—	on	on	12
—	—	—	on	8
on	—	—	on	9

You may want to vary the ti variable in the program depending on your PIC crystal speed. You can experiment with the ti variable until you find the best range for your particular PIC.

Troubleshooting

If the motor doesn't move at all, check the diodes. Make sure you have located them properly, facing in the direction shown in the schematic.

If the stepper motor moves slightly or quivers back and forth, it may be due to a number of possible causes:

- If using the battery power supply, the batteries may be too weak to power the motor properly. Note: Batteries wear out quickly because the current draw from the stepper motors is usually high.

- If you substituted the TIP120 NPN transistor for another transistor, the substitute transistor may not be switching properly or the current load of the stepper motor may be too great. Solution: Use TIP120 transistors.

- You have the stepper motor improperly wired into the circuit. Check the coils using an ohmmeter and rewire if necessary.

- The pulse frequency is too high. If the pulses to the stepper motor are going faster than the motor can react, the motor will malfunction. The pulse frequency is controlled by the ti variable in the program. Increasing the value of this variable will slow down the pulse frequency to the stepper motor. The solution to this is to reduce the pulse frequency.

UCN 5804 Dedicated Stepper Motor ICs

Until now, you have controlled the stepper motor directly from the PIC chip. However, dedicated ICs can be used to control the stepper motors directly. By incorporating step-

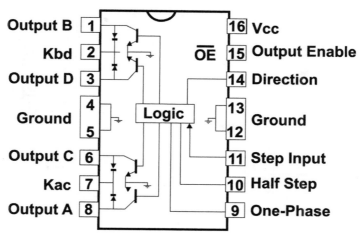

FIGURE 17-9   Pin from the UCN5804B IC

per motor controller chips into the design, the PIC controller can control multiple stepper motors. These controller chips can do most of the grunt work of controlling a stepper motor. This simplifies your program and overall circuit while enhancing the hardware, a good combination.

One chip I use frequently is the UCN5804 stepper motor controller. The pin from the UCN5804 is shown in Figure 17-9. The features of the UCN5804 are as follows:

- 1.25-ampere maximum output current (continuous)
- 35-volt output-sustaining voltage
- Full-step and half-step outputs
- Output Enable and direction control
- Internal clamp diodes
- Power-on reset
- Internal thermal shutdown circuitry

The schematic for the stepper motor controller using a dedicated IC is shown in Figure 17-10, and a photograph of the circuit is shown in Figure 17-11. The UCN5804 is powered by a 5-volt DC power supply. Although internally powered by 5 volts, it can control stepper motors voltages up to 35 volts.

Notice in the schematic that two resistors are labeled rx and ry and do not show any resistance value. Depending on the stepper motor, these resistors may not be necessary. Their purpose is to limit the current through the stepper motor to 1.25 amps (if necessary).

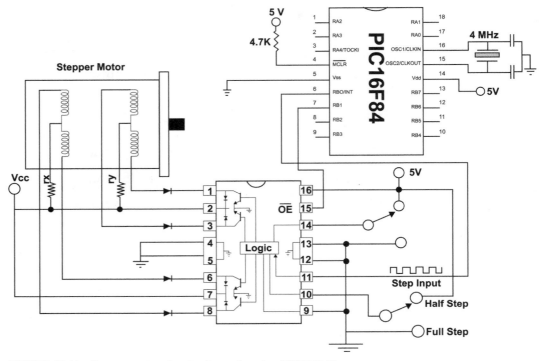

FIGURE 17-10   Stepper motor circuit schematic using UCN5804B

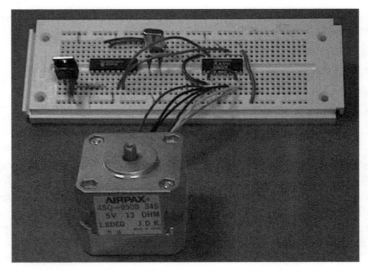

FIGURE 17-11   Stepper motor controller circuit

Let's look at your 5-volt stepper motor. It has a coil resistance of 13 ohms. The current draw of this motor will be .385 amps (5 volts/13 ohms) or 382 *milliamperes* (mA), well below the 1.25-amp maximum rating of the UCN5804. In this case, resistors rx and ry are not needed and may be eliminated from the schematic.

Before moving on, let's look at one more case involving a 12-volt stepper motor with a phase (coil) resistance of 6 ohms. The current drawn by this motor is 2 amps (12 volts/6 ohms), which is above the UCN5804 maximum current rating. To use this stepper motor, you must add the rx and ry resistors. The rx and ry resistor values should be equal to each other, so each phase will have the same torque. The values chosen for these resistors should limit the current drawn to 1.25 amps or less. In this case, the resistors should be at least 4 ohms (5 to 10 watts). With the resistor in place, the current drawn is 1.20 amps (12 volts/10 ohms).

The inputs to the UCN5804 are compatible with a *complementary metal oxide semiconductor* (CMOS) and *Time to Live* (TTL). In other words, you can connect the outputs from your PIC microcontroller directly to the UCN5804 and expect it to function properly. The Step Input (Pin 11) to the UCN5804 is generated by the PIC microcontroller. The Output Enable pin, when held low, enables the stepper motor; when brought high, it disables (stops) the stepper motor.

Pins 10 and 14 on the UCN5804 are controlled by switches that bring the pins to a logic high or low. Pin 10 controls whether the output to the stepper motor will be full-step or half-step and Pin 14 controls the direction. If you wanted, these options may also be put under the PIC control. The pins are brought to the logic high and low to activate the options just like the Output Enable pin.

The following is a PICBasic program that uses a dedicated stepper motor IC:

```
'PICBasic Stepper Motor w/UCN 5804 Program 17.5
Symbol TRISB = 134      'Initialize TRISB to 134
Symbol PortB = 6        'Initialize Port B to 6
Poke TRISB,0            'Set Port B lines output
low1                    'Bring Output Enable low to run
start:
pulsout 0, 10000        'Send 10 ms pulse to UCN5804
goto start              'Do again
```

The PICBasic Pro program is as follows:

```
'PICBasic Pro Stepper Motor w/UCN 5804 Program 17.6
TRISB = 0               'Set Port B lines output
low1                    'Bring Output Enable low to run
start:
pulsout 0, 10000        'Send 10 ms pulse to UCN5804
goto start              'Do again
```

In this case, I again wrote a simple core program to show how easy you can get the stepper motor running. You can, of course, add options to the program to change the pulse frequency, connect the direction and step mode pins, and so on.

TABLE 17-3   Additional components

1	UCN5804B stepper motor controller chip
1	Five-volt stepper motor unipolar (six-wire)
1	115-volt/5-volt step-down wall transformer
6	1N914 diodes
4	TIP120 NPN transistors
1	7805 voltage regulator
1	Rectifier (50-volt, 1 amp)
1	150 uF capacitor
1	4050 hex buffer chip

Parts List

The parts for this project are listed in Table 17-3 and in Chapter 1, "Microcontrollers." The components are available from the Images Company, Jameco Electronics, JDR Microdevices, and RadioShack (see the suppliers index).

Servomotors

Servomotors (see Figure 18-1) are used in radio-controlled model airplanes, cars, boats, and helicopters. Because of this hobbyist market, servomotors are readily available in a number of stock sizes. Servomotors have applications in animatronics, robotics, and positioning control systems.

Primarily, servomotors are geared DC motors with a positional feedback control that enables the rotor to be positioned accurately. The specifications state that the shaft can be positioned through a minimum of 90 degrees (+/−45 degrees). In reality, you can extend this range closer to 180 degrees (+/−90 degrees) by adjusting the positional control signal.

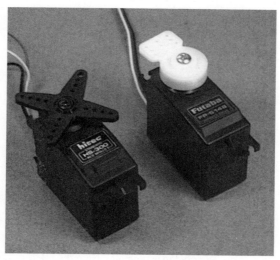

FIGURE 18-1   Picture of two 5-volt hobby servomotors with 42 oz. torque

A servomotor has three wire leads. Two leads are for power, +5V and GND, and the third lead feeds a position control signal to the motor. The positional control signal is a single variable-width pulse, which can be varied from 1 to 2 milliseconds. The width of the pulse controls the position of the servomotor shaft.

A 1-millisecond pulse rotates the shaft to the extreme *counterclockwise* (CCW) position (−45 degrees). A 1.5 millisecond pulse places the shaft in a neutral midpoint position (0 degrees). A 2-millisecond pulse rotates the shaft to the extreme *clockwise* (CW) position (+45 degrees).

The pulse width is sent to the servomotor approximately 50 times a second (50 Hz). Figure 18-2 illustrates the relationship of a pulse width to the servomotor position.

The first program sweeps the servomotor back and forth like a radar antenna. The schematic is shown in Figure 18-3. I'm keeping the program purposely small to illustrate the core programming needed. A picture of this project is shown in Figure 18-4.

The variable B3 holds the pulse width value. If you examine the pulsout command

```
Pulsout Pin, Period
```

Pin, of course, is the pin out. The number used in Period is specified in 10 microsecond (μsec) units. In the program, you are starting with B3 equaling 100, or 100 times 10 μsec, or 1 millisecond. If you look back at your servo specifications, you see in 1 millisecond the servo's arm rotate to its leftmost position.

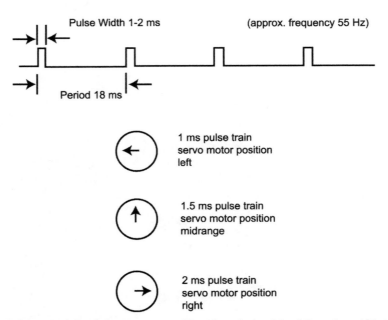

FIGURE 18-2   Pulse train delivered to servomotor. Note the relationship of the pulse width to the servomotor rotor's position.

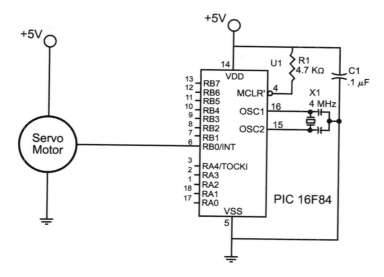

FIGURE 18-3 Schematic of basic servomotor sweeper controller (automatic)

FIGURE 18-4 Automatic servomotor sweeper

The program continues to smoothly increment the B3 variable, sweeping the servo motor's arm to its rightmost position at B3, which equals 200 (2 milliseconds). At this point, the process reverses and B3 begins to decrement back down to 100. This sweeping back and forth continues for as long as the program is run. The program code is as follows:

```
'PICBasic Servo Motor Program 18.1
'Sweep left to right like a radar antenna
b3 = 100                        'initialize at left position
sweep:
pulsout 0 ,b3                   'Send signal to servo motor
pause 18                        'transmit signal 50-60 Hz
b3 = b3 + 1                     'increment servo pulse width
if b3 > 200 then sweepback      'end of forward sweep?
goto sweep                      'keep sweeping
sweepback:
b3 = b3 - 1                     'decrement servo pulse width
pulsout 0, b3                   'send pulse to servo motor
pause 18                        'send it 50-60 Hz
if b3 < 100 then sweep          'end of sweepback?
goto sweepback                  'keep going back
```

The PICBasic Pro version is as follows:

```
'PICBasic Pro Servo Motor Program 18.2
'Sweep left to right like a radar antenna
b3 var byte
b3 = 100                        'initialize at left position
sweep:
pulsout 0 ,b3                   'Send signal to servo motor
pause 18                        'transmit signal 50-60 Hz
b3 = b3 + 1                     'increment servo pulse width
if b3 > 200 then sweepback      'end of forward sweep?
goto sweep                      'keep sweeping
sweepback:
b3 = b3 - 1                     'decrement servo pulse width
pulsout 0, b3                   'send pulse to servo motor
pause 18                        'send it 50-60 Hz
if b3 < 100 then sweep          'end of sweepback?
goto sweepback                  'keep going back
```

Extending Servo Motor Range

A pulse width variance from 1 to 2 milliseconds will provide a full 90 degrees of rotation. To extend this range up to 180 degrees, you need to use pulses smaller than 1 milliseconds and greater than 2 milliseconds.

If you decide to extend the rotational movement from your servo, you should be aware of certain problems that may arise. Primarily, the servomotor has end stops that limit how far the shaft can rotate in either direction. If the PIC is sending a signal to the servomotor that is past either end stop, the motor will continue to fight against the end stop. The servomotor in this stalled condition will draw an increased current and generate greater wear on the gearing inside the motor, neither of which is desirable.

An in-group variance exists among servomotors from the same manufacturer, as well as a variance among servomotor manufacturers. Although one servo may need a 2.8-millisecond pulse for a full rotation extension, another may only require a 2.4-millisecond pulse width.

When you decide to go out of the prescribed range of acceptable pulse widths (1 to 2 milliseconds) for servo motors, you should check the individual servomotor to ensure that you will not stall.

Manual Servo Control

This next project allows you to control the servomotor via a few switches. The schematic is shown in Figure 18-5. The switch used is a *single-pole, double-throw* (SPDT) with a center off position. The center off position is essential. Without it, you will need to use two switches instead of one.

The operation of this switch is simple. To activate the servo, push the switch in the upward direction and the servo will begin to rotate in one direction. To stop the motor, move the switch in the center off position. To rotate the servo in the opposite direction, push the switch lever down. Stop the servo as before by placing the switch in the center off position. The complete project is shown in Figure 18-6 and the program code is as follows:

```
'PICBasic Manual Servo Controller Program 18.3
Symbol porta = 5
b3 = 150                    'Initialize servo at center position
start:
peek porta,b0              'Look at switches on Port A
if bit0 = 0 then sweepl    'Is SW1 pressed?
if bit1 = 0 then sweepr    'Is SW2 pressed?
pulsout 0 ,b3              'Hold servo in current position
pause 18                   'Send signal 50-60 Hz
goto start                 'Check switches again
sweepl:                    'SW1 is pressed
b3 = b3 + 1                'Increment servo pulse width
pulsout 0 ,b3             'Send signal to servo motor
pause 18                   'Transmit signal 50-60 Hz
if b3 > 200 then hold1     'Maximum sweepl value?
goto start                 'Keep sweeping
sweepr:                    'SW2 is pressed
b3 = b3 - 1                'Decrement servo pulse width
pulsout 0, b3             'Send pulse to servo motor
pause 18                   'Send it 50-60 Hz
if b3 < 100 then hold2     'Minimum sweepr value?
goto start                 'Keep going back
hold1:'Hold maximum value 200
b3 = 200
goto start
hold2:'Hold Minimum value 100
b3 = 100
goto start
```

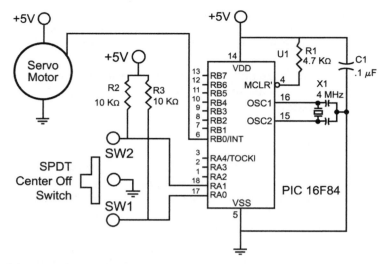

FIGURE 18-5   Schematic of manual control servo motor

FIGURE 18-6   Manual control servomotor circuit

The PICBasic Pro program is as follows:

```
'PICBasic Manual Servo Controller Program 18.4
b3 var byte
b3 = 150                        'initialize servo at center position
start:
'Peek porta to look at switches on Port A
if PortA.0 = 0 then sweep1    'Is SW1 pressed?
```

```
if PortA.1 = 0 then sweepr    'Is SW2 pressed?
pulsout 0 ,b3                 'Hold servo in current position
pause 18                      'send signal 50-60 Hz
goto start                   'Check switches again
sweepl:                      'SW1 is pressed
b3 = b3 + 1                  'increment servo pulse width
pulsout 0 ,b3                'Send signal to servo motor
pause 18                     'transmit signal 50-60 Hz
if b3 > 200 then hold1       'maximum sweepl value?
goto start                   'keep sweeping
sweepr:                      'SW2 is pressed
b3 = b3 - 1                  'decrement servo pulse width
pulsout 0, b3               'send pulse to servo motor
pause 18                     'send it 50-60 Hz
if b3 < 100 then hold2       'minimum sweepr value?
goto start                   'keep going back
hold1:'Hold maximum value 200
b3 = 200
goto start
hold2:'Hold Minimum value 100
b3 = 100
goto start
```

Multiple Servomotors

Using the routines in the last servomotor program, you can easily connect three servo-motors to the PIC 16F84 and still have four open *input/output* (I/O) lines available for other duties.

The next project provides manual control for multiple (two) servomotors. It uses the core programming routines used in the previous example. The schematic adds a switch and a servo motor (see Figure 18-7).

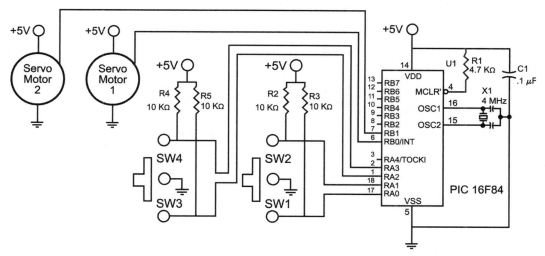

FIGURE 18-7   Schematic of multiple servomotor controller (manual)

```
'PICBasic Multiple Servo Controller Program 18.5
Symbol porta = 5
b3 = 150                        'initialize servo 1 at center position
b4 = 150                        'initialize servo 2 at center position
start:
peek porta,b0                   'Look at switches on Port A
if bit0 = 0 then sweepl         'Is SW1 pressed?
if bit1 = 0 then sweepr         'Is SW2 pressed?
if bit2 = 0 then sweepl2        'Is SW3 pressed?
if bit3 = 0 then sweepr2        'Is SW4 pressed?
pulsout 0 ,b3                   'Hold servo 1 in current position
pulsout 1, b4                   'Hold servo 2 in current position
pause 18                        'send signal 50-60 Hz
goto start                      'Check switches again
sweepl:                         'SW1 is pressed
b3 = b3 + 1                     'increment servo pulse width
pulsout 0 ,b3                   'Send signal to servo 1 motor
pulsout 1, b4                   'Hold servo 2 position
pause 18                        'transmit signal 50-60 Hz
if b3 > 200 then hold1          'maximum sweepl value?
goto start                      'keep sweeping
sweepr:                         'SW2 is pressed
b3 = b3 - 1                     'decrement servo pulse width
pulsout 0, b3                   'send pulse to servo motor
pulsout 1, b4                   'hold servo 2 position
pause 18                        'send it 50-60 Hz
if b3 < 100 then hold2          'minimum sweepr value?
goto start                      'keep going back
hold1:'Hold maximum value 200
b3 = 200
goto start
hold2:'Hold Minimum value 100
b3 = 100
goto start
'SECOND SERVO MOTOR ROUTINE-------------------------------------------------------
sweepl2:                        'SW3 is pressed
b4 = b4 + 1                     'increment servo pulse width
pulsout 1 ,b4                   'Send signal to servo 2 motor
pulsout 0, b3                   'Hold servo 1 position
pause 18                        'transmit signal 50-60 Hz
if b4 > 200 then hold3          'maximum sweepl value?
goto start                      'keep sweeping
sweepr2:                        'SW4 is pressed
b4 = b4 - 1                     'decrement servo pulse width
pulsout 1, b4                   'send pulse to servo 2 motor
pulsout 0, b3                   'hold servo 1 position
pause 18                        'send it 50-60 Hz
if b4 < 100 then hold4          'minimum sweepr value?
goto start                      'keep going back
hold3:'Hold maximum value 200
b4 = 200
goto start
hold4:'Hold Minimum value 100
b4 = 100
goto start
```

The PICBasic Pro program is displayed here:

```
'PICBasic Pro Multiple Servo Controller Program 18.6
b3 var byte
b4 var byte
b3 = 150                        'initialize servo 1 at center position
b4 = 150                        'initialize servo 2 at center position
start:
'Peek Port A to look at switches on Port A
if PortA.0 = 0 then sweepl      'Is SW1 pressed?
if PortA.1 = 0 then sweepr      'Is SW2 pressed?
if PortA.2 = 0 then sweepl2     'Is SW3 pressed?
if PortA.3 = 0 then sweepr2     'Is SW4 pressed?
pulsout 0 ,b3                   'Hold servo 1 in current position
pulsout 1, b4                   'Hold servo 2 in current position
pause 18                        'send signal 50-60 Hz
goto start                      'Check switches again
sweepl:                         'SW1 is pressed
b3 = b3 + 1                     'increment servo pulse width
pulsout 0 ,b3                   'Send signal to servo 1 motor
pulsout 1, b4                   'Hold servo 2 position
pause 18                        'transmit signal 50-60 Hz
if b3 > 200 then hold1          'maximum sweepl value?
goto start                      'keep sweeping
sweepr:                         'SW2 is pressed
b3 = b3 - 1                     'decrement servo pulse width
pulsout 0, b3                   'send pulse to servo motor
pulsout 1, b4                   'hold servo 2 position
pause 18                        'send it 50-60 Hz
if b3 < 100 then hold2          'minimum sweepr value?
goto start                      'keep going back
hold1:'Hold maximum value 200
b3 = 200
goto start
hold2:'Hold Minimum value 100
b3 = 100
goto start
'SECOND SERVO MOTOR ROUTINE-----------------------------------------------------
sweepl2:                        'SW3 is pressed
b4 = b4 + 1                     'increment servo pulse width
pulsout 1 ,b4                   'Send signal to servo 2 motor
pulsout 0, b3                   'Hold servo 1 position
pause 18                        'transmit signal 50-60 Hz
if b4 > 200 then hold3          'maximum sweepl value?
goto start                      'keep sweeping
sweepr2:                        'SW4 is pressed
b4 = b4 - 1                     'decrement servo pulse width
pulsout 1, b4                   'send pulse to servo 2 motor
pulsout 0, b3                   'hold servo 1 position
pause 18                        'send it 50-60 Hz
if b4 < 100 then hold4          'minimum sweepr value?
goto start                      'keep going back
hold3:'Hold maximum value 200
b4 = 200
goto start
hold4:'Hold Minimum value 100
b4 = 100
goto start
```

FIGURE 18-8   Multiple manual servomotor controller

The completed project is shown in Figure 18-8.

Timing and Servomotors

As you experiment with servomotors, it is necessary to feed the pulse signal to the servomotor 50 to 60 times a second. It is important to keep this in mind when running multiple servomotors or other time-critical applications.

PICBasic Pro Compiler Project: Five-Servomotor Controller

This next project is for the PIC Pro Basic compiler. The schematic shown in Figure 18-9 is a five-servomotor controller. This project may be purchased as a kit from Images SI, Inc. Figure 18-10 is a photograph of the five-servomotor controller kit assembled.

This project uses a larger 16F873 PIC microcontroller. When programming the 16F873, make sure the watchdog timer and the brownout reset are disabled in the EPIC Programmer Configuration Options menu. If the brownout reset is not disabled, the circuit may automatically reset whenever a servomotor draws enough current to make the supply voltage dip momentarily. This is not what you want to have happen in the middle of a robotic arm operation, so make sure that the configuration bit is disabled.

The PICBasic Pro program for the five-servomotor controller is as follows:

```
'PICBasic Pro Program 18.7
'Manual control of five servomotors using five SPDT switches
'Microcontroller PIC 16F873
adcon1 = 7      'Set Port A to digital I/O
```

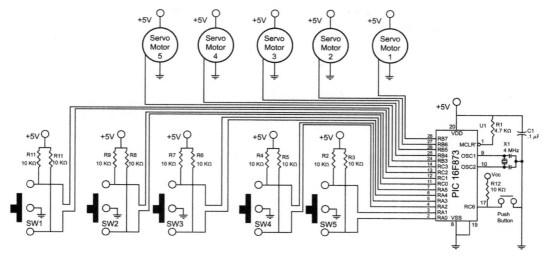

FIGURE 18-9 Schematic of five-servomotor controller

FIGURE 18-10 Five-servomotor kit assembled

```
'Declare Variables

B0 VAR BYTE      ' Use B0 as hold pulse width variable for servo 1
B1 VAR BYTE      ' Use B1 to hold pulse width variable for servo 2
B2 VAR BYTE      ' Use B2 to hold pulse width variable for servo 3
B3 VAR BYTE      ' Use B3 to hold pulse width variable for servo 4
B4 VAR BYTE      ' Use B4 to hold pulse width variable for servo 5
B6 VAR BYTE      'Variable for pause routine
B7 VAR WORD      'Variable for pause routine


'Initialize servomotor variables

B0 = 150         'start up position servo 1
B1 = 150         'start up position servo 2
B2 = 150         'start up position servo 3
```

```
B3 = 150        'start up position servo 4
B4 = 150        'start up position servo 5

start:

'Output servomotor position
PORTB = 0                           'Prevents potential signal inversion on reset
      PulsOut PORTB.7, B0           'send current servo 1 position out
      PulsOut PORTB.6, B1           'send current servo 2 position out
      PulsOut PORTB.5, B2           'send current servo 3 position out
      PulsOut PORTB.4, B3           'send current servo 4 position out
      PulsOut PORTB.3, B4           'send current servo 5 position out

'Routine to adjust pause value (nom 18) to generate approx. 50 Hz update

B7 = B0 + B1 + B2 + B3 + B4
B6 = B7/100
B7 = 15 - B6
Pause B7


' Check for switch closures

      IF PORTC.3 = 0 Then left1      'Is sw1 left active?
      IF PORTC.2 = 0 Then right1     'Is sw1 right active?
      IF PORTC.1 = 0 Then left2      'Is sw2 left active?
      IF PORTC.0 = 0 Then right2     'Is sw2 right active?
      IF PORTA.5 = 0 Then left3      'Is sw3 left active?
      IF PORTA.4 = 0 Then right3     'Is sw3 right active?
      IF PORTA.3 = 0 Then left4      'Is sw4 left active?
      IF PORTA.2 = 0 Then right4     'Is sw4 right active?
      IF PORTA.1 = 0 Then left5      'Is sw5 left active?
      IF PORTA.0 = 0 Then right5     'Is sw5 right active?

GoTo start


'Routines for Servomotor 1
left1:
      B0 = B0 + 1                    'increase the pulse width
      IF B0 > 254 Then max0          'maximum 2.54 milliseconds
      GoTo start
right1:
      B0 = B0 - 1                    'decrease the pulse width
      IF B0 < 75 Then min0           'minimum .75 milliseconds
      GoTo startp
max0:
      B0 = 254                       'cap max B1 at 2.54 milliseconds
      GoTo start
min0:
      B0 = 75                        'cap min B1 at .75 milliseconds
      GoTo start

'Routines for Servomotor 2
left2:
      B1 = B1 + 1                    'increase the pulse width
      IF B1 > 254 Then max1          'maximum 2.54 milliseconds
      GoTo start
right2:
B1 = B1 - 1                          'decrease the pulse width
```

```
    IF B1 < 75 Then min1            'minimum .75 milliseconds
    GoTo start
max1:
    B1 = 254                        'cap max B1 at 2.54 milliseconds
    GoTo start
min1:
    B1 = 75                         'cap min B1 at .75 milliseconds
    GoTo start

'Routines for Servomotor 3
left3:
    B2 = B2 + 1                     'increase the pulse width
    IF B2 > 254 Then max2           'maximum 2.54 milliseconds
    GoTo start
right3:
    B2 = B2 - 1                     'decrease the pulse width
    IF B2 < 75 Then min2            'minimum .75 milliseconds
    GoTo start
max2:
    B2 = 254                        'cap max B2 at 2.54 milliseconds
    GoTo start
min2:
    B2 = 75                         'cap min B2 at .75 milliseconds
    GoTo start

'Routines for Servomotor 4
left4:
    B3 = B3 + 1                     'increase the pulse width
    IF B3 > 254 Then max3           'maximum 2.54 milliseconds
    GoTo start
right4:
    B3 = B3 - 1                     'decrease the pulse width
    IF B3 < 75 Then min3            'minimum .75 milliseconds
    GoTo start
max3:
    B3 = 254                        'cap max B3 at 2.54 milliseconds
    GoTo start
min3:
    B3 = 75                         'cap min B3 at .75 milliseconds
    GoTo start

'Routines for Servomotor 5
left5:
    B4 = B4 + 1                     'increase the pulse width
    IF B4 > 254 Then max4           'maximum 2.54 milliseconds
    GoTo start
right5:
    B4 = B4 - 1                     'decrease the pulse width
    IF B4 < 75 Then min4            'minimum .75 milliseconds
    GoTo start
max4:
    B4 = 254                        'cap max B4 at 2.54 milliseconds
    GoTo start
min4:
    B4 = 75                         'cap min B4 at .75 milliseconds
    GoTo start

END
```

TABLE 18-1   Additional components for each servomotor

1	Four- to 6-volt HS 300 (42 oz. torque) servomotor or an equivalent
1	SPDT toggle switch with center off position
2	10 kilo-ohm, 1/4-watt resistors

TABLE 18-2   Additional components for five-servomotor controller

1	PIC16F873
1	4 MHz Xtal
2	22 pF capacitors
5	SPDT PC-mounted switches with center off position
11	10K, 1/4-watt resistors
1	4.7K resistor
1	.1 uf 50-volt capacitor
	Misc: 5-volt DC power supply

Parts List

In addition to the parts listed in Chapter 1, "Microcontrollers," the components listed in Table 18-1 would be necessary in this project for each servomotor. The parts list for the five-servomotor controller is displayed in Table 18-2. These parts are available from the Images Company, Jameco Electronics, JDR Microdevices, and RadioShack (see the suppliers index).

Controlling AC Appliances

In your previous work, you connected the output of your PIC microcontrollers to mostly *complementary metal oxide semiconductor* (CMOS) or *Time to Live* (TTL) logic (+5-volt) applications. What you will accomplish in this chapter is to use the PIC microcontroller output to control standard household *alternating current* (AC) loads and appliances.

For your demonstration project, you will take the toxic gas sensor from Chapter 15, "Analog-to-Digital (A/D) Converters," and make an automatic ventilation control system. This project is for demonstration purposes only. The PIC microcontroller reads the toxic gas sensor that senses the real-world environment and then, depending on that reading, controls a real-world appliance (a fan). It is *not*, under any circumstances, to be used in any commercial or private application as a toxic gas detector, controller, or ventilation control.

Your project is simple; when the PIC microcontroller senses a toxic gas, it will turn on an electric fan and keep it on until the gas sensor determines that the toxic gas concentration has returned to a safe level.

Before you build your project, you will need some basic information when designing an AC control system for yourself.

Inductive and Resistive Loads

Any device you power or control may be called a *load*. The electrical device (or load) will fall into one of two electrical categories: inductive or resistive. What type of device it is will determine the type of circuit used to power and control the load.

It's pretty easy to distinguish an inductive load from a resistive load. An inductive device (load) has wire coils and windings in it, such as motors, transformers, relays, and solenoids. A resistive device doesn't have any inductive coils or wire windings, such as incandescent lights, toasters, coffee makers, and heaters would.

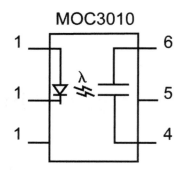

FIGURE 19-1   MOC 3010 pinout

To control AC loads, you will use an optocoupler triac. The MOC 3010 is such a device and is packaged as a six-pin dip chip (see Figure 19-1). When the PIC outputs a high signal (+5 volts) on one of its I/O pins connected to Pin 1 on the MOC3010, it will turn on an internal *light-emitting diode* (LED). The internal LED triggers a photosensitive internal triac (Pins 6 and 4) that in turn will trigger an external triac that powers the load.

Circuit Construction

To safely build these circuits, do *not* use the solderless breadboard. The voltages and currents are greater than what can be safely handled on the breadboard.

Please be careful; I don't want anyone accidentally shocking or electrocuting him- or herself. Always be extra careful when building any circuits that use household electric power. The power available from your household electric is more than enough to reduce your circuit to a cinder, give yourself a nasty shock, or worse.

Figure 19-2 shows a resistive-type AC appliance circuit fragment (minus the PIC controller). An inductive appliance controller is shown in Figure 19-3, again minus the PIC microcontroller. The resistor RL (in each schematic) is your main load or appliance that is being powered. The triac chosen will determine the maximum power of the appliance (in watts) that may be safely controlled. The power rating on the triac I used (see Table 19-1) is 6 amps at 200 volts, more than enough for a 50-watt fan.

I advise that you construct the inductive-type circuit because it can be used for both resistive and inductive types of appliances. This will alleviate the question later on if the device (load) is resistive or inductive. The schematic fragment for resistive loads can be used for a comparison with the inductive circuit or, if you want, as dedicated resistive load controllers.

Since most readers will be interested in controlled AC appliances or devices in their home, Figure 19-4 is the circuit you will build. All the components must be soldered to a *printed circuit board* (PCB). Make sure any lines and wiring carrying the household power are adequately insulated and covered.

Resistive Load

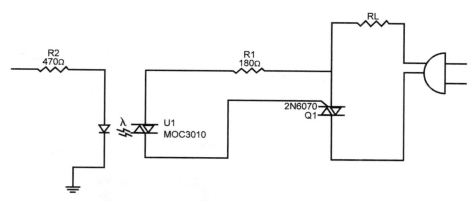

FIGURE 19-2   Resistive AC appliance circuit fragment

Inductive Load

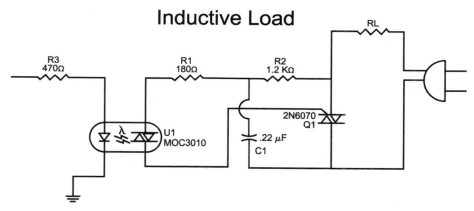

FIGURE 19-3   Inductive AC appliance circuit fragment

The triac I used is rated at 200 volts at 6 amps, which means that it is capable of handling 1,200 watts. In order to pass that much current, the triac requires a substantial heat sink. I advise you to keep the maximum power under 250 watts. The program code is as follows:

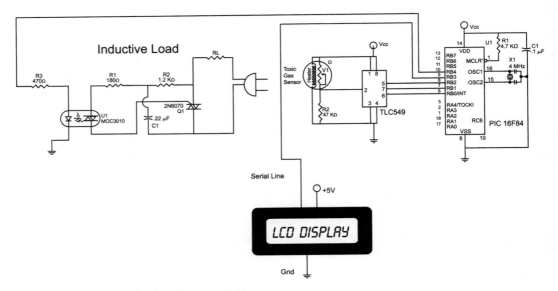

FIGURE 19-4 Schematic of appliance controller

```
'PICBasic Serial A/D Converter & Toxic Gas Program 19.1
low 1                      'Bring I/O clock low
start:
gosub serial_in
'LCD Routine
serout 3, N2400, (254, 1)
pause 1
serout 3, N2400, (#b0)
pause 100                  'Let me see display
if b0 > 190 then fan1      'Turn fan on
if b0 < 191 then fan2      'Turn fan off
goto start
'Serial In Routine
serial_in:
low 2                      'Bring CS down low
bit7 = pin0                'Load bit7 into B0
pulsout 1,1                'Bring clk pin high then low
bit6 = pin0                'Load bit6 into B0
pulsout 1,1                'Bring clk pin high then low
bit5 = pin0                'Load bit5 into B0
pulsout 1,1                'Bring clk pin high then low
bit4 = pin0                'Load bit4 into B0
pulsout 1,1                'Bring clk pin high then low
bit3 = pin0                'Load bit3 into B0
pulsout 1,1                'Bring clk pin high then low
bit2 = pin0                'Load bit2 into B0
pulsout 1,1                'Bring clk pin high then low
```

```
bit1 = pin0                 'Load bit1 into B0
pulsout 1,1                 'Bring clk pin high then low
bit0 = pin0                 'Load bit0 into B0
pulsout 1,1
high 2                      'Bring CS high
return
fan1:
high 4
goto start
fan2:
low 4
goto start
```

The code for the PICBasic Pro program is as follows:

```
'PICBasic Pro Serial A/D Converter & Toxic Gas Program 19.2
b0 var byte
low 1                       'Bring I/O clock low
start:
gosub serial_in
' LCD Routine
serout 3, 4, [254, 1]
pause 1
serout 3, 4, [#b0]
pause 100                   'Let me see display
if b0 > 190 then fan1       'Turn fan on
if b0 < 191 then fan2       'Turn fan off
goto start
'Serial In Routine
serial_in:
low 2                       'Bring CS down low
b0.7 = pin0                 'Load bit7 into B0
pulsout 1,1                 'Bring clk pin high then low
b0.6 = pin0                 'Load bit6 into B0
pulsout 1,1                 'Bring clk pin high then low
b0.5 = pin0                 'Load bit5 into B0
pulsout 1,1                 'Bring clk pin high then low
b0.4 = pin0                 'Load bit4 into B0
pulsout 1,1                 'Bring clk pin high then low
b0.3 = pin0                 'Load bit3 into B0
pulsout 1,1                 'Bring clk pin high then low
b0.2 = pin0                 'Load bit2 into B0
pulsout 1,1                 'Bring clk pin high then low
b0.1 = pin0                 'Load bit1 into B0
pulsout 1,1                 'Bring clk pin high then low
b0.0 = pin0                 'Load bit0 into B0
pulsout 1,1
high 2                      'Bring CS high
return
fan1:
high 4
goto start
fan2:
low 4
goto start
```

Test Circuit

If you want to test the program and circuit without building the entire circuit, you can. Remove the appliance power control section and replace it with a resistor and an LED (see Figure 19-5). When the microcontroller senses toxic gas, it turns on the LED. When the gas dissipates, the LED goes off.

Smart Control

Smart controls need feedback to determine if a particular action is being performed. To make this point, I wish to draw an analogy.

Let's say you've just returned from a newspaper stand with your favorite magazine. You sit in a chair, turn on a lamp, and, behold, no light. "Darn," you say to yourself. You look down to the socket to see if the lamp is plugged in, which it is. You look over to the clock on the wall that's on the same fuse as the lamp. The clock is ticking away, so you

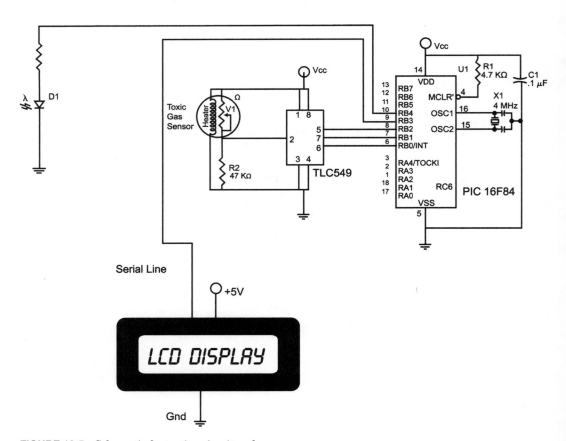

FIGURE 19-5 Schematic for testing circuit and program

know you have juice going to the lamp. You flick the lamp switch a couple of times to make sure the switch isn't stuck. Now you take the lampshade off the lamp, and sure enough a black spot on the bulb lets you know that it's burnt out. You replace the bulb, the lamp works fine, and you finally get to read the magazine.

There's nothing remarkable about this incident, but it is a good example of a smart control. When the switch was turned on but the lamp didn't light, you went through the various steps to locate and correct the problem. But what about a microcontroller? Had it been the microcontroller's job to turn on the lamp, it would not know whether or not the light was on.

To build a smart control, you must give the microcontroller feedback to check to see if the action was successful. For the light example, you might use a photocell or photoresistor for a feedback signal. If the feedback gave a negative response, the microcontroller could ring an alarm or send a signal.

Keep this information in mind so if you find someday that you have a need for a smart controller somewhere, your PIC microcontroller can handle the job.

Electronic Nose

The toxic gas sensor is the forerunner of the electronic nose. High-quality, highly selective electronic noses have numerous applications in many industries, such as food storage, the medical field, and law enforcement.

The toxic gas sensors are not digital sensors; they are analog. If sufficient in-group variance and response to compounds take place, a group of sensors can be wired into a neural network for specific order detection.

Parts List

Along with the components listed in Chapter 1, "Microcontrollers," Table 19-1 lists the other necessary parts. Also needed here are a toxic gas sensor and an *analog-to-digital* (A/D) converter (refer to Chapter 15).

These parts are available from the Images Company, James Electronics, JDR Microdevices, and RadioShack (see the suppliers index).

TABLE 19-1 Additional components

Item	RadioShack PN #
MOC3010 Optocouple triac	PN 276-134
2N6070 Triac (or equivalent)	PN 276-1000
.22 uF capacitor	PN 272-1070
Line cord	PN 278-1255
470-ohm, 1/4-watt resistor	PN 271-1317
180-ohm, 1/4-watt resistor	PN RSU 11344702
1.2-kilo-ohm, 1/4-watt resistor	PN RSU 11344884
AC small fan	

A Few More Projects

In this chapter, you will outline a few more projects for the PICBasic Pro compiler. These projects show the versatility of what you can accomplish.

Binary Clock

Your first project is a binary clock. The reason I am calling it a binary clock is that the seconds, minutes, and hours are displayed as discrete rows of binary numbers. You will use *light-emitting diodes* (LEDs) connected to Port B to generate the display. The schematic for the clock is shown in Figure 20-1. Notice in schematic that the Port B lines from PB0 to PB6 are connected to and shared by the LEDs for the seconds, minutes, and hours. The program and circuit multiplexes the display.

Look at the transistors at the bottom of each LED display. Each transistor controls the LED display in its column. When the transistor is turned on, it connects that LED column to ground, allowing the lines on Port B to light the LEDs. When the transistor is turned off, signals from Port B cannot light any LEDs. The 16F84 microcontroller selects each row of LEDs in sequence by turning the bottom transistors on and off. The code for the program is as follows:

```
'Binary Clock - PICBasic Pro Program 20.1
TRISB = 0                  'Port B = output
TRISA = 15
PORTA = 0

bs VAR BYTE                'Variable to hold seconds
bm VAR BYTE                'Variable to hold minutes
bh VAR BYTE                'Variable to hold hours
ct VAR WORD                'Loop variable

Start:                     'Increment second
bs = bs + 1
If bs = 61 Then
```

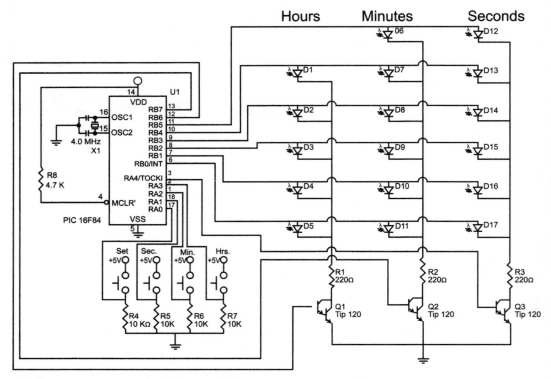

FIGURE 20-1 Binary display clock

```
      bs = 0
      bm = bm + 1
EndIF
If bm = 61 Then            'Check minutes
      bm = 0
      bh = bh + 1
EndIF
If bh = 25 Then            'Check hours
      bh = 0
EndIF
'Display clock info for 1 second
For ct = 0 to 321
PORTB = bs                 'Place second's # on Port B
PORTA.4 = 1                'Display seconds
Pause 1                    'Pause for a millisecond
PORTA.4 = 0                'Turn off seconds display
PORTB = bm                 'Place minutes # on Port B
PORTB.7 = 1                'Display minutes
Pause 1                    'Pause a millisecond
PORTB.7 = 0                'Turn off minute display
PORTB = bh                 'Place hours # on port b
```

```
PORTB.6 = 1                'Display hours
Pause 1                    'Pause a millisecond
PORTB.6 = 0                'Turn off hours display
Next ct
If PORTA.0 = 1 Then set
GoTo Start
set:
If PORTA.1 = 1 Then        'Increment seconds
     bs = bs + 1
     PORTB = bs
     PORTA.4 = 1           'Display seconds
     Pause 500             'Don't change numbers too fast
     PORTA.4 = 0           'Turn off display
EndIF
If bs > 60 Then bs = 0     'Error checking
If PORTA.2 = 1 Then        'increment minutes
     bm = bm + 1
     PORTB = bm
     PORTB.7 = 1           'Display minutes
     Pause 500             'Don't change numbers too fast
     PORTB.7 = 0           'Turn off display
EndIF
If bm > 60 Then bm = 0     'Error checking
If PORTA.3 = 1 Then        'Increment hours
     bh = bh + 1
     PORTB = bh
     PORTB.6 = 1           'Display hours
     Pause 500             'Don't change numbers too fast
     PORTB.6 = 0           'Turn off display
EndIF
If bh > 24 Then bh = 0     'Error checking
If PORTA.0 = 1 Then set
GoTo Start
```

Setting the Clock

To set the clock, press the Set switch connected to RA0. While the set switch is pressed, the clock is in Set mode. Pressing the second (Sec.) switch connected to RA1 will increment the seconds. Pressing the minute (Min.) switch connected to RA2 will increment the minutes. Pressing the hours (Hrs.) switch will increment the hours. Releasing the Set switch will start the clock again with the new setting.

The way the clock is programmed, it displays the hours in a 24-hour format. This can be changed to a 12-hour format.

The PIC 16F84 was not made to be a clock. The tests I have been able to run show the binary clock to be accurate to a few seconds per day. This is not the fault of the microcontroller but in my ability to accurately measure the time elapse. Using a faster 16F84 chip and Xtal (20 MHz) could help generate a clock with great accuracy. If I develop a method to increase the accuracy of the binary clock, it will be posted on www.imagesco.com.

Nothing is critical about the construction. This is the type of project that begs for a PC board to keep the wiring of the LEDs accurate and neat. By the time this book goes to press, this project will be a kit offered by the Images Company.

Digital Geiger Counter

Images SI, Inc. sells an inexpensive GCK-03 or GCK-04 Geiger counter with a digital output (see Figure 20-2). The digital output is provided from a ⅛-inch jack on the left-hand side of the Geiger counter. You can read this digital output using your 16F84 and provide the counts per second and give an approximate radiation reading in millirads per hour.

The schematic for the circuit is shown in Figure 20-3. The pulses are read from the ⅛-inch plug inserted into the Geiger counter.

The program for the counter is provided here. With some simple modification, the program can be made to output to a serial *liquid crystal display* (LCD).

```
'Digital Geiger_Counter -- *** PICBasic Pro Program 20.2
'Count Pulses on Port B.0
'Display mR/h on LCD display
'Declare variables
w1 VAR WORD
w2 VAR WORD


b1 VAR BYTE
b2 VAR BYTE
b3 VAR BYTE
```

FIGURE 20-2   Geiger counter with digital output

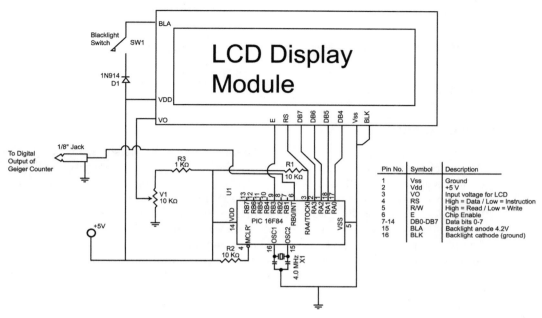

FIGURE 20-3 Digital Geiger counter schematic

```
'Main Routine


start:


Count PORTB.0, 1000, w1      'Count pulses every 1 second


LCDOut 254,128               'Bring LCD cursor to line 1 first position
LCDOut "Counts/Sec ", #w1," "

If W1 <= 20 Then
     W2 = W1 * 5
EndIF
If W1 >20 Then
     W1 = W1 * 63
     w2 = div32 10
EndIF

w1 = w2

b1 = w1 DIG 2
b2 = w1 DIG 1
b3 = w1 DIG 0
```

```
LCDOut $fe,$c0                    'Bring LCD cursor to line 2
LCDOut "mR/h " , #b1 ,".",#b2,#b3," "

'Clear the counting variable
w1 = 0

'Do it Again
GoTo start
```

Frequency Generator

This next project is a square wave frequency generator. The square wave produced is a 50 percent duty cycle. You can change the frequencies by opening and closing the switches connected to Port A (RA0 to RA4). In closing the switches, you are creating a binary number on Port A; the program reads the number and then outputs a frequency. Table 20-1 provides the frequencies for the first 16 binary numbers using a 4.0 MHz crystal. The schematic for the circuit is shown in Figure 20-4.

Higher frequencies can be obtained by using a faster PIC microcontroller. A few other options that would not be too hard to implement would be varying the duty cycle of the square wave and possibly a one-shot and/or a timer. The frequency generator code is as follows:

```
'Frequency Generator - PICBasic Pro Program 20.3
TI VAR WORD
Delay VAR WORD
TRISB = 128
PORTB.0 = 0
```

TABLE 20-1   Binary Number Frequencies

Binary number	Output frequency
1	17.8 kHz
2	9.6 kHz
3	6.5 kHz
4	5.0 kHz
5	4.0 kHz
6	3.4 kHz
7	2.9 kHz
8	2.5 kHz
9	2.3 kHz
10	2.0 kHz
11	1.85 kHz
12	1.7 kHz
13	1.58 kHz
14	1.47 kHz
15	1.34 kHz
16	1.29 kHz

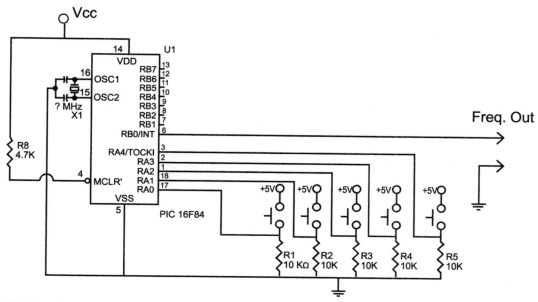

FIGURE 20-4   Frequency generator schematic

```
'Read Port A switches

Start:
TI = PORTA
If TI = 0 Then TI = 1
Delay = TI * 24

If PORTB.7 = 0 Then start
run:
PORTB.0 = 1
Pauseus Delay
PORTB.0 = 0
Pauseus Delay
If PORTB.7 = 1 Then run
GoTo Start
```

In Closing

All the programs in this book are freely available at www.imagesco.com for you to download.

Suppliers Index

Images SI Inc.
109 Woods of Arden Road
Staten Island, NY 10312
718-966-3694 telephone
718-966-3695 fax

Jameco Electronics
1355 Shoreway Road
Belmont, CA 94002-4100
650-592-8097
650 -592-2503

JDR Microdevices
1850 South 10th Street
San Jose, CA 95112-4108
1-800-538-500

Radio-Shack
(see local listings)

Hexadecimal Numbers

The hexadecimal number system is a base-16 system. It was found early on that base-16 hexadecimal is an ideal number system for working with digital computers. This is because each nybble (4-bit number) is also base 16. Thus, one hexadecimal digit can represent a 4-bit number.

As computer processors have grown, they have used larger and larger numbers. Regardless how large these numbers become, they are for the most part evenly divisible by four bits. Early personal computers were 8-bit and then came 16-bit processors; current personal computers are 32-bit. Future computers will be 64, then 128, and so on. The PIC microchips do not follow this convention and use an odd number of bits throughout their internal structure (refer to Figure 6-6). See Table A-1.

Notice that 10 in hex is equal to 16 in your standard base 10. Sometimes a lowercase h is placed behind the number to identify it as a hex number and not base 10. The letters A through F stand for hexidecimal digits 10 through 15. An 8-bit number can be represented by two hexadecimal digits. The hex number FF represents the maximum number a 8-bit digit can hold, 255.

TABLE A-1   Decimal, binary, and hexadecimal numbers

Decimal	Binary	Hexadecimal
0	0000	0
1	0001	1
2	0010	2
3	0011	3
4	0100	4
5	0101	5
6	0110	6
7	0111	7
8	1000	8
9	1001	9
10	1010	A
11	1011	B
12	1100	C
13	1101	D
14	1110	E
15	1111	F
16	00010000	10h

As with base 10, each number going toward the left is a multiple. For instance, 20 in base 10 is really 2 times 10. In hexadecimal, 20 is 2 times 16.

Many compilers and assemblers use hexadecimal numbers to program code. When examining code for the PIC microcontrollers, you will undoubtedly run across hexadecimal as well as binary numbers.

Index

Note: Boldface numbers indicate illustrations.

B

C

E

T

U

V